Inorganic Fibres
and
Composite Materials

EPO APPLIED TECHNOLOGY SERIES
VOLUME 3

The surveys of this series have been made by the examiners of the European Patent Office at The Hague (The Netherlands) who are most competent in the technology concerned.

The present survey encompasses the most recent developments in the field and rests on a selection made by the authors among the patent and non-patent literature available at the EUROPEAN PATENT OFFICE (EPO).

EPO APPLIED TECHNOLOGY SERIES

OTHER TITLES IN THE SERIES

Pergamon Journal of Related Interest *(free specimen copy gladly sent on request)*

World Patent Information — The international journal for patent information and industrial innovation

PERGAMON INTERNATIONAL INFORMATION CORPORATION

Pergamon International Information Corporation (PIIC), located in McLean, Virginia, USA, develops and acquires electronic databases for on-line dissemination through the Pergamon InfoLine computer service. PIIC also markets the Pergamon InfoLine service in North America.

PIIC's VIDEO PATSEARCH[R] system is a revolutionary, computerized, search system that displays both patent text and drawings utilizing interactive laser videodisc display. This system, which displays both patent drawings and chemical structures, is the first effective in-house patent search system.

VIDEO PATSEARCH[R] is based on PATSEARCH[R], the database of US and patent cooperation treaty patents development by PIIC which is available for direct computer access through the Pergamon InfoLine service. Other patent related products of PIIC include the International Patent Documentation Center (INPADOC) worldwide patent database, PCT patent publications and a patent search center in the US Patent and Trademark Office.

PERGAMON INTERNATIONAL INFORMATION CORPORATION
1340 Old Chain Bridge Road
McLean, VA 22101
USA

Inorganic Fibres and Composite Materials

A Survey of Recent Developments

P. BRACKE, H. SCHURMANS

and

J. VERHOEST

European Patent Office, The Hague, The Netherlands

Pergamon International Information Corporation
a member of the Pergamon Group
PERGAMON PRESS

OXFORD · NEW YORK · TORONTO · SYDNEY · PARIS · FRANKFURT

U.K.	Pergamon Press Ltd., Headington Hill Hall, Oxford OX3 0BW, England
U.S.A.	Pergamon Press Inc., Maxwell House, Fairview Park, Elmsford, New York 10523, U.S.A.
	Pergamon International Information Corporation, 1340 Old Chain Bridge Road, McLean, VA 22101, USA
CANADA	Pergamon Press Canada Ltd., Suite 104, 150 Consumers Road, Willowdale, Ontario M2J 1P9, Canada
AUSTRALIA	Pergamon Press (Aust.) Pty. Ltd., P.O. Box 544, Potts Point, N.S.W. 2011, Australia
FRANCE	Pergamon Press SARL, 24 rue des Ecoles, 75240 Paris, Cedex 05, France
FEDERAL REPUBLIC OF GERMANY	Pergamon Press GmbH, Hammerweg 6, D-6242 Kronberg-Taunus, Federal Republic of Germany

First edition 1984

Library of Congress Cataloging in Publication Data
Bracke, P.
Inorganic fibres & composite materials.
(EPO applied technology series ; v. 3)
1. Inorganic fibers.   2. Fibrous composites.
I. Schurmans, H.   II. Verhoest, J.   III. Title.
IV. Title: Inorganic fibres and composite materials.
V. Series.
TS1549.A1B73   1984   666'.8   83-25736

British Library Cataloguing in Publication Data
Bracke, P.
Inorganic fibres & composite materials.—
(EPO applied technology series; no.3)
1. Textile fibers, synthetic
I. Title   II. Schurmans, H.
III. Verhoest, J.   IV. Series
677'.4   TS1548.5
ISBN 0-08-031145-8

In order to make this volume available as economically and as rapidly as possible the author's typescript has been reproduced in its original form. This method unfortunately has its typographical limitations but it is hoped that they in no way distract the reader.

Printed in Great Britain by A. Wheaton & Co. Ltd., Exeter

Preface

This monograph aims to present a timely summary of the recent developments
in the field of inorganic fibres and composite materials, as it emerges from
the published patent and non-patent literature incorporated in the systematic
documentation of the EPO at The Hague (Netherlands).
Said study covers in principle the period from 1970 up to now, but as an
introduction to each chapter a comprehensive but concise survey of the state
of the art prior to that period is provided, accompanied by ample though not
exhaustive references to patents, periodicals or books.
Where possible an indication of the expected trend concludes each chapter.
Although not limited to that country, special attention has been given to
progress achieved in Japan.

The monograph contains two major parts; a first one dealing with the
composition, preparation and specific treatment of the fibres, while a second
part concerns their use in the manufacture of high grade composite materials.

Part one embraces, in principle, all types of artificial inorganic fibres
with the exception however of glass fibres; due to their specific nature,
properties and methods of manufacture, glass fibres are part of a quite
distinct technological field and should be dealt with in a separate
monograph.
On the other hand, monocrystalline fibres (e.g. whiskers) are included in the
present study; development of that type of fibre competed during the last
decade with the progress in continuous filament technology, especially for
obtaining composite materials with exceptional characteristics.
Due to their particular methods of preparation they are treated in a
separate chapter of Part one.

Specific entries are furthermore provided for the most important types of
fibres known up to date; for each type the following aspects are to be
discussed : manufacture, post-treatments, properties and uses.

Part two is limited to entirely inorganic materials, *i.e.* the matrix as well
as the reinforcing fibres being inorganic. An extension to the extremely vast
field of the fibre reinforced resins would increase the size of the present
monograph out of proportion and seriously affect its clarity.
Fibre reinforced plastics should thus constitute the subject of a separate
monograph.

More detailed information on the scope of a particular item will be given in
said item.

This study has been made for the Commission of the European Communities, DG XIII - Information Market and Innovation, Division Technological Information and Patents.

Contents

PART II   INORGANIC FIBRES COMPOSITE MATERIALS

Note on Cited Patent Documents

In this monograph, a great number of patents and published patent applications* are cited, using an international two-letter country code, i.e.:

        DE = Germany (Federal Republic)
        EP = European Patent Office
        FR = France
        GB = United Kingdom
        JP = Japan
            [all cited documents** are
             published patent applications
             (KOKAI TOKKYO)]
        SU = USSR
        US = United States of America
        WO = International Bureau of W.I.P.O.
            (PCT-applications)

Whenever corresponding patent documents have been found, they are indicated preceeded by an "=" sign.

---

* The patent literature covered by the search files of the European Patent Office at The Hague (Netherlands) encompasses patent publications of following countries or offices : Australia, Austria, Canada, France, Germany (Federal Republic), Japan, Switzerland, United Kingdom, USA, USSR, European Patent Office (EPO), World Intellectual Property Organization (W.I.P.O.), African Intellectual Property Organisation (OAPI), Belgium, Luxembourg, The Netherlands.

** The *KOKAI TOKKYO's* numbers begin actually with two digits indicating the year of publication *in the Japanese system*; we have replaced that indication by the occidental (gregorian) year; the Japanese year equal to the gregorian *minus* 25.

Example : Japanese # 52. 84122 is cited as 77. 84122.

PART I

INORGANIC FIBRES THEIR MANUFACTURE AND PROPERTIES

NOTES ON CONTENTS

This part covers artificially made, inorganic fibres with the exception of glass fibres.

Only these processes are considered that permit the production of fibrous material that can be recovered as such. In-situ growing of fibres in matrices is dealt with in Part II, page 126.

All fibres showing a monocrystalline structure are included in Chapter 7 irrespective of their nature or chemical composition.

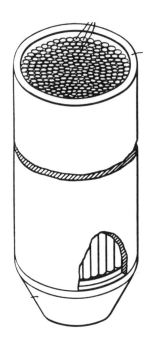

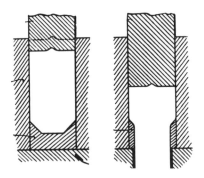

Obtaining small diameter filaments by drawing matrix encased preforms

CHAPTER 1

Metal Fibres

1. Summary of the Prior Art

Metal fibres which are to be used as reinforcing elements in composite
materials should possess specific mechanical and physical properties such as
high tensile strength and elasticity modulus, low density, appropriate thermal
expansion coefficient and stability against dissolution by, or chemical
reaction with, the matrix material. Such properties are chiefly determined
by the composition and the metallurgical structure of the fibre material,
the latter being in turn governed by the manufacturing process and the
thermal or thermomechanical treatment of the wire product.
The present chapter will only deal with the development of the relevant
manufacturing processes.

Among the many metals which have been investigated as to their strengthening
ability in different matrix materials, only filaments of the refractory
metals titanium, tungsten, tantalum and molybdenum and filaments of beryllium,
steel and some superalloys were found to have an acceptable combination of
the required properties. These filaments are generally fabricated by one of
the following techniques:

1.1. Wire drawing techniques.   This conventional, well known method has
been used for the production of fibres of nearly all of the above mentioned
metals and proved to be satisfactory. But production costs as well as
problems of wire breakage, which rise very rapidly with decreasing filament
thickness, are the limiting factors in the production of filaments or fibres
with diameters below 100µ.

Smaller diameters down to 10µ or even less have been obtained by encasing a
core wire within a sheath or matrix of a ductile sacrificial material, drawing
the whole to a predetermined cross section and removing the sheath material
by etching. This technique is advantageously used for the simultaneous
drawing of a plurality of filaments (1).

In an analogous embodiment metal material was sheated with a glass or
ceramic envelope, heated to a temperature sufficient to soften the sheath
material and to melt or soften the metal core and drawing the whole while in
plastic state. By using high frequency heating very fine wires of high
melting metals could be obtained (2)(3)(4).

1.2. Melt forming techniques.   Metal melts have a very low viscosity which
excludes a priori direct fibre drawing from the melt. On the other hand their
relatively high surface energy which often exceeds the viscosity, makes the
production of filaments by extrusion or casting of a melt very problematic

Melt Forming Techniques

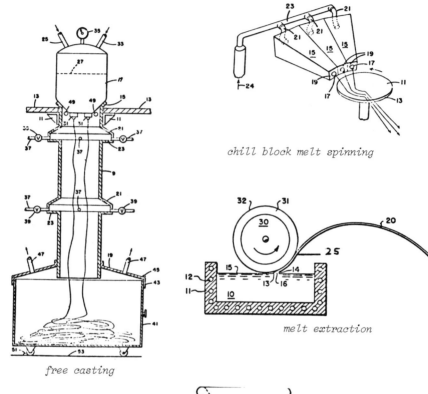

free casting

chill block melt spinning

melt extraction

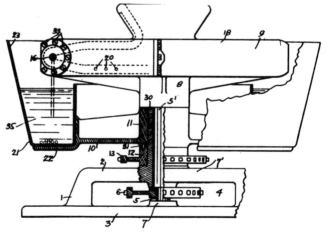

centrifugal spinning

as the filamentary jet, as soon as it issues from the shaping die, easily
breaks up into separate droplets before sufficient heat can be withdrawn for
its solidification. It was therefore necessary to augment the extrusion speed
by gas pressure or centrifugal force in such a way as to maintain the
filamentary shape of the jet long enough to stabilize it by some means.
In application of this principle, different methods have been developed.
According to a first embodiment called chill roll casting or chill block
melt spinning the extruded molten jet impinges on a moving chill surface
such as an endless band or a rotating drum, where it almost instantaneously
solidifies. Wire thickness can be varied by varying the extrusion pressure,
the moving speed of the chill surface, temperature of the melt, etc. (5)(6).
Filaments thus obtained usually present a flattened cross section.
A variant of this method, called melt extraction, involves the use of a
rotating heat-extracting wheel in contact with the surface of a source of
molten metal. The metal film in contact with the wheel solidifies, adheres
thereto and is so extracted from the melt and subsequently released.
This method was especially used for the production of fibres of low melting
metals, such as zinc, lead, magnesium, etc. (23).

Another method used the idea of the centrifugal casting whereby the molten
metal contained in a rotating crucible was projected through a series of
openings (7)(8). This method normally does not allow the production of
continuous filaments.

In the "free casting" technique the molten jet is extruded into a cooling
fluid where it solidifies before coming in contact with a solid surface (9).
But due to the inviscid character of the metal melt, process parameters
are very critical which renders this technique rather troublesome.
An improvement of this method consisted of extruding the molten metal into
some atmosphere reactive to the molten metal thus forming a protective and
stabilizing skin on the molten jet (10)(11).

1.3. Powder metallurgical methods.   Metal powders possibly mixed with some
binder are extruded at ambient or high temperature through an appropriate
die and then further compacted and/or sintered.
These methods are especially suitable for metals which are difficult to
process by reason of their high melting point, brittlenes or reactivity e.g.
the refractory metals (12).
They have however only rarely been used for the production of fine wires (13).

2. Developments since 1970

No fundamentally new processes for the production of metal fibres have been
reported since 1970.

In the wire drawing techniques, the manufacture of fine filaments by drawing
sheated or matrix-inbedded wires has been further improved by the use of
matrix materials of different compositions or shape, special dies, etc.
It must be noted that since 1970 also Japanese industry has shown an increased interest in this technique (14)(15)(16)(17)(18).
A particular variant of said technique is called the "in situ" drawing of
fibres and involves the extrusion or drawing of a compacted and possibly
sintered or remelted preform composed of a blend of different metal powders.
The fibres can be recovered by chemical dissolving of the matrix material.
Although said method has effectively been used for the production of
discrete needlelike fibres (35), it is especially suited for the direct
manufacture of metal/metal composites (36) (see also part II).

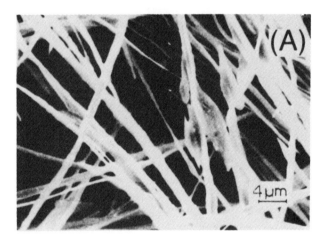

*Powder metallurgical fabrication of Niobium fibres obtained
by "in-situ drawing" and removal of a copper matrix*

More activity could be observed in the field of the melt forming techniques.
The chill roll melt spinning was adapted by ALLIED CHEMICAL CORP. for the
production of filaments or ribbons composed of amorphous alloys which proved
to have exceptional mechanical properties (19)(20)(22) and whose use as
reinforcing fibres for composite materials has already been suggested (21).
ALLIED CHEMICALS has now virtually monopolized the field of the amorphous
alloys.

The melt extraction technique also gained renewed interest. More accurate
control of process parameters and the use of more sophisticated apparatus
now permit the production of filaments of high melting metals such as
zirconium, steel, etc. (24)(25)(26)(27)(28).
The "free casting" method involving skin formation on the extruded jet was
further improved by MONSANTO (29)(30)(31)(32) and also by MICHELIN who uses
the method for the production of steel cord for tires (33)(34).

*   *

*

REFERENCES TO CHAPTER 1

(1)   US 3394213   (ROEHR PRODUCTS)
(2)   US 3214805   (DUPONT DE NEMOURS)
(3)   US 3362803   (W. DANNOEHL)
(4)   FR 1423604   (W. DANNOEHL)
(5)   US 2886866   (MARVALAUD INC.)
(6)   FR 1168521   (MARVALAUD INC.)
(7)   US 3466352   (CORBETT ASSOC.)
(8)   GB 1136732   (BRUNSWICK CORP.)
(9)   US 2976590   (MARVALAUD)
(10) US 3216076   (CLEVITE CORP.)
(11) NL 6604168   (MONSANTO)
(12) US 3264388   (KAISER ALUMINIUM CHEMICAL CORP.)
(13) US 3199331   (NAT. RESEARCH CORP.)
(14) US 3838488   (SUMITOMO ELECTRIC IND.)
(15) NL 7112396   (NIPPON SEISEN)
(16) US 3643304   (NIPPON SEISEN)
(17) DE 2339466   (NIPPON SEISEN)
(18) FR 2192882   (NIPPON SEISEN)
(19) US 3862658   (ALLIED CHEMICAL CORP.)
(20) US 4260007   (ALLIED CHEMICAL CORP.)
(21) FR 2281434   (ALLIED CHEMICAL CORP.)
(22) US 4331739   (ALLIED CHEMICAL CORP.)
(23) US 1879336   (F.W. FOLEY)
(24) US 3843762   (G. SLEIGH)
(25) US 3871439   (BATTELLE DEVELOPMENT CORP.)
(26) US 4150708   (GTE SYLVANIA)
(27) US 4157729   (GTE SYLVANIA)
(28) US 4259125   (RIBBON TECHNOLOGY)
(29) FR 2111009   (MONSANTO)
(30) US 3720741   (MONSANTO)
(31) US 3727292   (MONSANTO)
(32) US 3854518   (MONSANTO)
(33) FR 2393635   (MICHELIN)
(34) FR 2367564   (MICHELIN)
(35) J.P. LUCAS *et al.*, "Journ. of the Am. Ceramic Soc." vol 63, n° 5 - 6 (May-June 1980) p 280
(36) J.H. SWISHER *et al.* "Journ. of Applied Physics", vol 41 n° 3 (March 1970) p 1097 - 1098

CHAPTER 2

Carbon Fibres

1. Summary of the Prior Art

Carbon textile materials such as yarn, felt, cloth, etc. have been made for a long time; their methods of manufacture nearly always involve a carbonisation step of an organic precursor normally present in the required textile form; but the products obtained so far exhibited only moderate strength values and found industrial application as heating elements, filter material, for thermal insulation, etc.

The present chapter will more specifically deal with the manufacture, properties, etc. of high strength and high modulus carbon fibres required to obtain composite materials of exceptionnally high specific stiffness and strength.

Extensive investigation carried out during the 1960 had revealed that strength and modulus of a carbon fibre are a function of its crystal structure; the ideal carbon fibre having a graphite structure with the graphite crystals oriented parallel to the fibre axis and having virtually no porosity between the individual fibrils.

Numerous techniques have been proposed to approximate as close as possible said ideal structure; most of these techniques belong to one of the 3 basic process features which are :

(1) Selection of the precursor material.
Adequate precursor materials should provide a high carbon yield and a low weight loss during carbonisation resulting in reduced shrinkage.
They preferably should also present a high degree of molecular orientation that can be maintained or even improved during the subsequent processing.

(2) Pretreatments previous to the carbonisation step.
These treatments are primarily intended to stabilize the fibre to prevent melting or deterioration during carbonisation and normally involve a heating step enhancing some molecular rearrangements such as cyclization, crosslinking, etc. Occasionally it gives rise to sixmembered rings from which will arise graphite.
Advantageously these treatments are combined with some mechanical action (e.g. stretching) to maintain or improve molecular orientation.

(3) Selection of process conditions during the carbonisation and/or graphitisation stages.
Said conditions have a profound effect on the mechanical properties of the carbon fibre, which may be varied over a relatively wide range by suitable

Micrograph of a highly oriented carbon fibre

choice of said conditions. Especially the internal relationship between tensile strength and tensile modulus will be primarily defined by the process parameters selected for the carbonisation and graphitisation.

Up to 1970 very extensive research on carbon fibres has been carried out in Japan, USA, UK and France and a tremendous number of publications have been issued in that period.
Here below are listed the industrial firms and research centres which for a substantial part contributed to the industrial development of carbon fibres.

JAPAN : JAPANESE BUREAU OF INDUSTRIAL TECHNICS;
        TOKAI DENKYOKU SEIZO; KUREHA;
        TORAY INDUSTRIES; JAPAN EXLAN.

UK   : ROYAL AIRCRAFT ESTABLISHMENT; ROLLS-ROYCE;
        COURTAULDS.

USA   : UNION CARBIDE; CELANESE CORPORATION; MONSANTO;
        GREAT LAKES CARBON.

The most promising techniques for the manufacture of high strength carbon fibres which were available at the start of the last decade are summarised below :

- The cellulose precursor process :
  this involved a preheating step, a carbonisation step and finally a graphitising step under hot stretching. Said hot working appeared to be a vital improvement as it gave rise to a remarkably better alignment of the graphite crystals and a reduced porosity.
  Different methods and/or apparatus to apply controlled amounts of stretch, were developed. Also the use of a catalyst to reduce overall process time was proposed by different investigators.

- The polyacrylonitrile (PAN) precursor process:
  the PAN precursor appeared to be advantageous compared to rayon as it presented already a high degree of molecular orientation. The process comprised generally different stages, such as :

  1° A heating stage in inert atmosphere or vacuum to cause cyclisation of the PAN molecular chain structure.

  2° Oxidation in air at temperatures between 250 - 400° C resulting in crosslinking reaction between molecular chains.
  Preferably the oxidation step is accompanied by some stretching action.

  3° Carbonisation.

  4° Graphitisation.

Besides the above mentioned processes, many others which started from precursor materials such as wool, lignin, pitch, polyvinylalcohol, polyvinylchloride, etc. have been investigated.

2. Developments since 1970

2.1. Manufacture processes.   Investigations on high strength carbon fibre manufacture from cellulose still continued after 1970 but little or no substantial progress has been made in that field since then.
Most efforts were indeed concentrated on improving the PAN precursor process which became the most competitive process in the field.
More recently new horizons were opened with the discovery of a new low cost

precursor material derived from pitch. New developments made with these three basic processes will now be discussed more in detail hereafter.

2.1.1. Cellulose (rayon) precursor process.   Most of the investigations aimed at the reduction of the carbonisation time and at the attaining of higher carbon yields thus rendering the process industrially more attractive. To that purpose it was proposed to impregnate the cellulose precursor with specific carbonisation catalysts :

- I. STEVENS & CO proposed organic phosphorus compounds (esters, phosphonic acids, etc.) (1)(2).
- THE CARBORUNDUM CO reported successive treatments with silicon compounds ($SiCl_4$) and nitrogen compounds (3).
- US SECRETARY OF THE NAVY LAB. described treatment with organic phosphorus-boron compounds (4).

Other provisions were intended to prevent deterioration of the cellulose yarn during the initial pyrolysis stage, which was manifested by a complete loss of strength, by impregnation of the precursor with strength increasing agents :

- WHITE E.F.T. proposed treatment with a combination of urea or its derivatives and ammonium phosphates (5).
- NITTO BOSEKI describes the use of sulfur oxy-anion compounds which should react with the cellulose (6)(7).
- US ATOMIC ENERGY CO suggested the impregnation with polyethylene/propylene (8).

A study of the more recent literature shows however that for the manufacture of high strength fibres the cellulose precursor process has been abandonned in favour of other more attractive methods. Cellulose on the other hand is now increasingly used for the manufacture of carbon textile materials useful for industrial applications other than composite materials, such as activated carbon fibres, cloth, etc. (9)(10).

2.1.2. PAN precursor process.   Much effort was invested in the improvement of the process which has now become the classical industrial way for the manufacture of high strength carbon fibres.
The improvements related to a large range of aspects of the PAN process; they will be discussed below in function of the most important aspects.

(i) Selection and preparation of specific precursors

- CELANESE CORP. reported improved spinning processes producing yarns with highly fibrillous and dense structure (11)(12)(13).
- TORAY IND. started from an interlaced non twisted yarn presenting no flaws nor cracks (14).
  A higher carbon yield could be obtained by using a copolymer precursor of acrylonitrile and an aromatic substituted hydroxymethyl acryl compound (15).
- JAPAN EXLAN CO described different pretreatments of PAN fibres prior to further processing such as reducing its water content (16), stretching in hot acid (17) or in hot water (18).
  Improving the structural soundness (in particular absence of flaws) also resulted in higher quality products (19).

- NATIONAL RESEARCH DEVELOPMENT reported improved copolymer compositions such as :

    acrylonitrile-vinyl alcohol or hydroxy ethylene (20)
    acrylonitrile-vinylchloride-itaconic acid (21)

- KANEGAFUCHI BOSEKI proposed a chemical transformation of the PAN polymer material into polyacrylamidoxium permitting shorter carbonisation time (22).

(ii) Cyclisation of the PAN precursor

Different new techniques were proposed to enhance the cyclisation reactions in the PAN molecules.

- CELANESE used different Lewis acid compounds, incorporated directly in the precursor material, or used as surface treatment agents (23)(24) (25)(26)(27)(28).
- MONSANTO proposed stabilisation of the PAN fibre by treatment in a catalyst containing hot polyol solution (29)(30).
- SIGRI ELECTRO GRAPHIT GmbH reported a heat treatment in solutions containing carboxylic acids, sulfonic acids or phenols (31).
- JAPAN EXLAN proposed treatment in hot nitrophenol solutions (32).

(iii) Oxidation stage

This is considered the most decisive step as it will to a large extent govern the final molecular orientation of the fibre and hence its ultimate strength properties. Much effort has been focused on improving said step. Acceleration of the oxidation was sought through the addition of specific catalysts.

- CELANESE reported the impregnation of the PAN fibres with solutions of persulfate (33), cobalt salts (34), an iron (II)/hydrogen peroxyde combination (35) or with acids (36).

A more accurate control of the exothermic oxidation reactions was obtained by impregnating the fibre or adding to the precursor composition flame retarders or antioxidants, thus avoiding local melting, collapsing or fluffing of the yarn.

- W. TURNER proposed hydroxylamine containing solutions (37).
- CELANESE reported aminophenol quinones (38) or aminosiloxanes (39).
- JAPAN EXLAN used primary amines or quaternary ammonium salts (40).
- MINNESOTA MINING AND MANUF. CO described impregnation with amines at high pressure and temperature (41).

Oxidation of the yarn in atmospheres other than air was proposed by some investigators

- TORAY INDUSTRIES reported the oxidation in nitrous oxide (42).
- MONSANTO : oxidation in a mixture of Br_2 and O_2 (43).
- UNION CARBIDE : oxidation in HCl and O_2 containing atmosphere (44).

cyclisation

oxidation

carbonisation

Molecular Rearrangements in the PAN precursor during the process steps

As the prior art had already revealed the importance of a stretching action
during the oxidation, different tentatives were made to take full advantage
of this phenomenon :

- MITSUBISHI RAYON reported stretching a PAN-vinylcopolymer fibre containing
  a copper catalyst up to 80 times its original length (45).
- FIBER MATERIALS INC. impregnated the fibre with a carboxylic acid capable
  of delivering its anhydride when heated, and oxidized under tension (46).
- Different companies such as GREAT LAKES CARBON CORP. (47), COURTAULDS LTD
  (48), BAYER (49), NATION RESEARCH (50)(51), UNION CARBIDE (44), JAPAN
  EXLAN (52) and SECURICUM (53)(54) suggested to split up the oxidation
  step into a so called preoxidation and a real oxidation while maintaining
  the yarn under different tensions.

(iv) Carbonisation

Prior to carbonisation, the oxidized PAN yarn can be submitted to some
intermediate treatments such as drying to eliminate all traces of humidity
(55)(56)(60), cleaning to remove absorbed oxygen (57) or tar products (58)
or impregnating with protecting agents (59)(60) permitting the fast
carbonisation. The carbonisation which is normally conducted in neutral or
reducing atmospheres can be improved by heating in acid and/or water vapor
(61)(62) or atmospheres containing small amounts of oxygen.

(v) Graphitisation

The main problem inherent to this step is to find an optimal time-temperature
related heating schedule that gives rise to maximum strength and modulus
value. Adequate schedules have been reported by CELANESE (63) and GENERAL
ELECTRIC (64), while TORAY mentioned multi-step graphitisation (65)(66).
On the other hand the graphitisation process can be accelerated in the
presence of catalyst such as boron compounds (67)(68).

2.1.3. Pitch precursor process.   Pitch had already been proposed in the
prior art as a low cost, easily available precursor material suitable to
replace the rather expensive polyacrylonitrile.
More recently new precursor materials based on blends of pitches and resins
have been suggested by KUREHA K.K.K. (69), CHARBONNAGES DE FRANCE (70)(71)
(72), CELANESE (73) and the CARBORUNDUM CO (74). But carbon fibres obtained
from pitches exhibited in general inferior strength properties compared to
the PAN process fibres.
A major break-through in the use of pitch came with the discovery in the
late sixties of the *mesophase* which constitutes the higher molecular weigth
insoluble portion of a pitch and which can be described as clusters of large,
predominantly aromatic molecules arranged in parallel. It was found that
some pitches having high mesophase content could easily be transformed by
extrusion or spinning into a precursor fibre with exceptional molecular
orientation that could be further processed into a high strength graphite
fibre. Raising the mesophase content of pitches was therefor one of the
primary goals and different processes to achieve this have been decribed
since 1970.

- L.I. GRINDSTAFF *e.a.* started from petroleum coal- or ethylene tar pitch
  which was heated at temperatures above 400°C until 75% mesophase has been
  formed (75).
- UNION CARBIDE improved said process by passing an inert gas through the
  pitch during formation of the mesophase (76)(77) or by subjecting it to

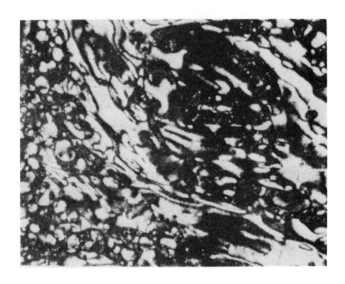

Pitch with high mesophase content

reduced pressure (78)(79). More recently pure physical processes such as solvent extraction, devolatilisation, etc. were also proposed (80).

- KUREHA K.K.K. obtained a high mesophase pitch by subjecting tetrabenzophenazine to a heat treatment at 450 - 600°C (81). In another embodiment a petroleum pitch was subjected to a two step heating (82).

- EXXON RESEARCH reported solvent extraction techniques to isolate the mesophase (83)(84)(85)(86).

- TOA NENRYO K.K. separated the mesophase by heat treatment of a pitch at a temperature above 380°C followed by allowing the mesophase to coalesce and settle (87)(88).

- FUJI STANDARD RESEARCH INC. described a process to prepare a pitch with "dormant" anistropic character by partially hydrogenating the mesophase. When applying shearing forces to the pitch, the anisotropic character reappeared (89).

The further processing of the pitch fibre for its transformation into carbon or graphite fibre is very analogous to the PAN fibre process. Prior to carbonisation the pitch fibre must also undergo some stabilisation to avoid melting or collapsing; said stabilisation generally comprises an oxidation step in air. It is also possible to oxidise the pitch before spinning (90)(91). Other possibilities are the oxidation in an atmosphere of Cl_2 and O_2 or in Cl_2, Br_2 and HNO_3 containing solutions (92)(93).

2.1.4. Other processes.   Besides the above mentioned most important precursor materials, $i.e.$ rayon/polyacrylonitrile/pitch, many other polymeric substances have been investigated since 1970 for their suitability as precursor material for the manufacture of high strength carbon fibre.

Examples of these precursors are:

▶Polymerisation resins

polybutadiene	UBE (94),
polyacetylene	GENERAL ELECTRIC (95)(96),
vinyl polymers	WACKER CHEMIE (97), DOW CHEMICAL (98),
	KUREHA (99), UNION CARBIDE (100),
polyethylene, polystyrene	SUMITOMO (101)(102).

▶Polycondensation resins

polyamides	ROLLS-ROYCE (103),
polyaramides	MONSANTO (104)(105), ONERA (106),
polyurethanes	CHEMOTRONICS (107),
polybenzimidazoles	CELANESE (108).

Finally, it could be interesting to mention a rather exotic carbon fibre manufacture process involving a thermal vapor phase decomposition of hydrocarbons such as benzene; carbon fibres with lengths up to 25 cm have been obtained (109)(110)(111).

2.2. Post-treatments of carbon fibres.   Research on special post-treatments of carbon fibres started in the late sixties, $i.e.$ as soon as the fibres became industrially available and the first attempts to make composites revealed unexpected problems, in particular failing bond strength and reduced shear resistance.

Surface defect on a carbonized PAN fibre

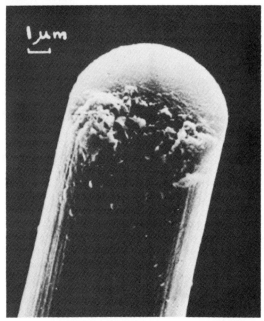

Fibre fracture caused by surface defect

So most of the treatments were intended to modify the surface characteristics of the carbon fibre prior to its incorporation in a matrix material, and aimed more specifically at improvement of the adherence between fibre and matrix material, and reduction of the oxidation rate and/or reaction rate with the matrix material.
Among the numerous techniques which have been proposed since, two substantially different kinds can be distinguished, namely :

- treatments consisting of a surface activation without having any detectable coating

- treatments resulting in a permanent coating of the fibre.

(i) Surface activation

Such treatments normally involve some etching or roughening of the fibre surface in order to increase its specific surface area and thus provide improved adherence. As a supplementary advantage, etching also contributes to the strength behaviour of the fibre itself by eliminating surface defects such as flaws, cracks, etc.
Various methods and chemicals have been suggested to achieve that goal but the major problem was to find an optimal combination of reagents and process parameters that would lead to an effective activation without any appreciable deterioration of the strength characteristics of the fibre.
The most representative publications on this subject are summarized below :

▶ Activation with gaseous reagents
.. in oxygen, ozone or water-containing atmospheres
   CELANESE CORP. (112)(113)
   UNITED AIRCRAFT CORP. (114)(115)
   LOCKHEED CORP. (116)
   UNITED ATOMIC ENERGY AUTHORITY (117)
.. in nitrogen or ammonia-containing atmospheres
   KUREHA KAGAHU K.K.K. (118)
   MONSANTO (119)
   GREAT LAKES CARBON CORP. (120)
.. in halogen-containing atmospheres
   UNION CARBIDE (121)
   COURTAULDS (122)
.. in sulfur oxyde-containing atmospheres
   TORAY IND. (123)

▶ Activation with liquid reagents or solutions; these normally consist of or contain oxidants such as
.. nitric acid UNITED AIRCRAFT (124)
.. halogens, halogen oxyacids
   UNITED AIRCRAFT (125)(126)
   V.R. DEITZ (127)
   GREAT LAKES CARBON CORP. (129)
   MORGANITE MODMOR LTD (130)
.. chromates, metalsalts ...
   GREAT LAKES CARBON CORP. (131)
   MONSANTO (132)
   GENERAL ELECTRIC (133)

Surface activation by electrochemical methods was suggested by HERCULES (134), COURTAULDS (135), RHONE-POULENC (136) and TOHO BESLON (137), while CELANESE utilised plasma or high radio frequency heating (138)(139).

(ii) Coatings

▶   With inorganic materials

Such coatings are frequently applied on fibres that are to be imbedded in inorganic matrices; their primary object is to confer to the fibre a better resistance against oxidation during the high temperature processing of the composite material or to inhibit any chemical reaction between fibre and matrix material that could give rise to fibre degradation.
Different coating materials have been proposed :

.. carbon coatings; these are mostly amorphous and are obtained by chemical
   vapor decomposition of hydrocarbons :
   DOW CHEMICAL CO (140)(141),
   DUNLOP (142),
   NATIONAL RESEARCH DEVELOPMENT CORP (143),
.. coatings of carbides, borides, nitrides :
   M.S. RASHID (144),
   UNION CARBIDE (145),
   THOMPSON FIBERGLASS (146),
   AVCO (147),
.. metal coatings : used to improve the wettability of the fibres by
   molten matrix material :
   ROLLS ROYCE (148),
   ELBAN (149),
   CELANESE (150),
   R.T. PEPPER (151),
.. glass or ceramic coatings
   silicondioxide :
   HAWKINS (152),
   aluminium phosphates :
   IMPERIAL CHEMICAL INDUSTRIES (153)(154),
   glass :
   ENGLISH ELECTRIC (156).

▶   With organic materials

Carbon fibres exhibit an inherent lack of wettability versus resins and this phenomenon was mainly responsible for the low shear strength of many carbon-resin composites.
Another problem was linked with the relative stiffness of the carbon fibre which rendered weaving or braiding troublesome.
These problems could be,at least partially,overcome by prewetting the fibres with suitable resin compositions such as :

.. epoxy resins
   MITSUBISHI RAYON (157), TOHO BESLON (157),
   RHONE-PROGIL (159), HERCULES (160),
   GREAT LAKES CARBON (161);

.. polyolefins
   ROLLS ROYCE (162), BAYER (163);

.. fluorinated polymers
   J.H. ROSS (164);

.. polyurethanes
   UNITED AIRCRAFT (165);

.. polyphenylenes :
   CELANESE (166)(167)
.. novolac resins :
   GREAT LAKES CARBON (168),
   CELANESE (169)
.. polyimides:
   TRW (170)

Besides the above mentioned treatments which are specific for the manufacture
of composite materials, other techniques have been reported to confer
particular properties to carbon fibres.

Although in some way these fall outside the scope of the present monograph, it
could be interesting to cite by way of example the following patent
publications which are related to the following items :

- improving the electrical conductivity of the fibres :
  CELANESE (171),
  F. VOGEL (172).

- confering ion exchange properties to the carbon surface by sulfonation,
  carboxylation, etc. :
  MITSUBISHI RAYON (173),
  KUREHA (174),
  ERMULENKO *et al.* (175),
  SECRETARY OF STATE OF DEFENSE (176).

- improving absorptive properties for active carbon fibres :
  LONZA-WERKE (177).

                        *   *

                             *

REFERENCES TO CHAPTER 2

(1)   US 3527564   (J. STEVENS)
(2)   US 3617220   (J. STEVENS)
(3)   US 3689220   (THE CARBORUNDUM CO)
(4)   US 3859043   (THE SECRETARY OF THE NAVY)
(5)   EP 7693   (E.F.T. WHITE)
(6)   US 3639140   (NITTO BOSEKI)
(7)   US 3661616   (NITTO BOSEKI)
(8)   US 3607672   (US ATOMIC ENERGY COMMISSION)
(9)   GB 1505095   (CLAIRAIRE LTD)
(10)   US 3976746   (HITCO)
(11)   US 3657409   (CELANESE)
(12)   US 3846833   (CELANESE)
(13)   US 3841079   (CELANESE)
(14)   FR 2393087   (TORAY IND.)
(15)   DE 2042358   (TORAY IND.)
(16)   DE 2506344   (JAPAN EXLAN)
(17)   GB 1500675   (JAPAN EXLAN)
(18)   GB 2011364   (JAPAN EXLAN)
(19)   FR 2236034   (JAPAN EXLAN)
(20)   FR 2328787   (NATIONAL RESEARCH DEVELOPMENT CORP.)
(21)   FR 2328723   (NATIONAL RESEARCH DEVELOPMENT CORP.)
(22)   DE 2158798   (KANEGAFUCHI BOSEKI)
(23)   US 3592595   (CELANESE)
(24)   US 3647770   (CELANESE)
(25)   US 3813219   (CELANESE)
(26)   US 3729549   (CELANESE)
(27)   US 4002426   (CELANESE)
(28)   FR 2076203   (CELANESE)
(29)   FR 2107609   (MONSANTO)
(30)   US 3814377   (MONSANTO)
(31)   DE 2220614   (SIGRI ELEKTROGRAPHIT GmBH)
(32)   DE 2361190   (JAPAN EXLAN)
(33)   US 3650668   (CELANESE)
(34)   US 3656882   (CELANESE)
(35)   US 3656883   (CELANESE)
(36)   US 3708326   (CELANESE)
(37)   US 3767773   (W. TURNER)
(38)   US 4004053   (CELANESE)
(39)   US 4009248   (CELANESE)
(40)   US 4024227   (JAPAN EXLAN)
(41)   US 4031288   (MINNESOTA MINING AND MANUFACTURE CO)
(42)   FR 2073796   (TORAY IND)
(43)   DE 2013913   (MONSANTO)
(44)   FR 2207088   (UNION CARBIDE)

(45)	US 3917776	(MITSUBISHI RAYON)
(46)	GB 2084978	(FIBER MATERIALS INC)
(47)	FR 2056271	(GREAT LAKES CARBON CORP.)
(48)	FR 2087946	(COURTAULDS)
(49)	FR 2175882	(BAYER)
(50)	FR 2204570	(NATIONAL RESEARCH)
(51)	DE 2045680	(NATIONAL RESEARCH)
(52)	GB 1499457	(JAPAN EXLAN)
(53)	GB 2014971	(SEGURICUM)
(54)	US 4100004	(SEGURICUM)
(55)	US 3677705	(CELANESE)
(56)	GB 1340069	(NATIONAL RESEARCH DEVELOPMENT CORP.)
(57)	DE 2420101	(JAPAN EXLAN)
(58)	DE 2407372	(JAPAN EXLAN)
(59)	US 3656903	(CELANESE)
(60)	US 4039341	(NATIONAL RESEARCH DEVELOPMENT CORP.)
(61)	US 3972984	(NIPPON CARBON CO)
(62)	FR 2084597	(BAYER)
(63)	US 3900556	(CELANESE)
(64)	US 3764662	(GENERAL ELECTRIC)
(65)	US 4301136	(TORAY IND)
(66)	EP 24277	(TORAY IND)
(67)	US 3723605	(CELANESE)
(68)	FR 2022221	(NATIONAL RESEARCH DEVELOPMENT CORP.)
(69)	US 3639953	(KUREHA KAGAGU KKK)
(70)	FR 2159660	(CHARBONNAGES DE FRANCE)
(71)	FR 2087413	(CHARBONNAGES DE FRANCE)
(72)	US 3966887	(CHARBONNAGES DE FRANCE)
(73)	US 4020145	(CELANESE)
(74)	US 3903220	(CARBORUNDUM CO)
(75)	US 3787541	(L.I. GRINDSTAFF)
(76)	US 3974264	(UNION CARBIDE)
(77)	EP 44761	(UNION CARBIDE)
(78)	US 3995024	(UNION CARBIDE)
(79)	US 4303631	(UNION CARBIDE)
(80)	EP 26647	(UNION CARBIDE)
(81)	US 4016247	(KUREHA KAGAGU KKK)
(82)	US 4115527	(KUREHA KAGAGU KKK)
(83)	US 4277325	(EXXON RESEARCH)
(84)	FR 2396793	(EXXON RESEARCH)
(85)	FR 2453886	(EXXON RESEARCH)
(86)	EP 34910	(EXXON RESEARCH)
(87)	EP 44714	(TOA NENRYO K.K.)
(88)	EP 55024	(TOA NENRYO K.K.)
(89)	EP 54437	(FUJI)
(90)	FR 2135128	(KOPPERS)
(91)	FR 2219906	(KOPPERS)
(92)	FR 2296032	(UNION CARBIDE)
(93)	FR 2118974	(COAL INDUSTRIES)
(94)	DE 2421443	(UBE)
(95)	US 3899574	(GENERAL ELECTRIC)
(96)	DE 2208212	(GENERAL ELECTRIC)
(97)	GB 1177739	(WACKER CHEMIE)
(98)	US 3840649	(DOW CHEMICAL)
(99)	US 3666417	(KUREHA K.K.K.)
(100)	US 3488151	(UNION CARBIDE)

(101)   FR 2216227   (SUMITOMO)
(102)   FR 2174951   (SUMITOMO)
(103)   FR 1603812   (ROLLS-ROYCE)
(104)   FR 2017523   (MONSANTO)
(105)   GB 1236282   (MONSANTO)
(106   FR 2411256   (OFFICE NATIONAL D'ETUDES ET DE RECHERCHE AEROSPATIALE)
(107)   DE 2559608   (CHEMOTRONICS)
(108)   US 3903248   (CELANESE)
(109)   T. KOYAMA $e.a.$ Jap. j. of Applied Physics vol 11 n°4 p 445-449
(110)   M. ENDO   Jap. j. of Applied Physics vol 15 n°11 p 2073-2076
(111)   M. ENDO   Jap. j. of Applied Physics vol 16 n° 9 p 1519-1523
(112)   US 3723607   (CELANESE)
(113)   US 3754957   (CELANESE)
(114)   US 3720536   (UNITED AIRCRAFT CORP.)
(115)   US 3772429   (UNITED AIRCRAFT CORP.)
(116)   US 3816598   (LOCKHEED AIRCRAFT)
(117)   FR 1600655   (U.K. ATOMIC ENERGY AUTHORITY)
(118)   JA 7222679   (KUREHA KKKK)
(119)   US 3627466   (MONSANTO)
(120)   US 3776829   (GREAT LAKES CARBON)
(121)   FR 2123366   (UNION CARBIDE)
(122)   GB 1379547   (COURTAULDS)
(123)   JA 7224977   (TORAY IND)
(124)   US 3660140   (UNITED AIRCRAFT)
(125)   US 3772350   (UNITED AIRCRAFT)
(126)   US 3801351   (UNITED AIRCRAFT)
(127)   US 3931392   (VR. DEITZ)

(129)   US 3746560   (GREAT LAKES CARBON)
(130)   FR 2084811   (MORGANITE MODMUR)
(131)   US 3989802   (GREAT LAKES CARBON)
(132)   US 3627570   (MONSANTO)
(133)   US 3870444   (GENERAL ELECTRIC)
(134)   US 3832297   (HERCULES)
(135)   FR 2039709   (COURTAULDS)
(136)   FR 2186972   (RHONE-POULENC)
(137)   GB 2071702   (TOHO BESLON)
(138)   US 3767774   (CELANESE)
(139)   US 3745104   (CELANESE)
(140)   US 3853610   (DOW CHEMICAL)
(141)   US 3908061   (DOW CHEMICAL)
(142)   FR 2062169   (DUNLOP)
(143)   FR 2098508   (NAT. RESEARCH DEVELOPMENT CORP.)
(144)   US 3023029   (M.S. RASHID)
(145)   FR 2192067   (UNION CARBIDE)
(146)   DE 1469488   (THOMPSON FIBER GLASS CO)
(147)   GB 2081695   (AVCO)
(148)   FR 2022113   (ROLLS-ROYCE)
(149)   US 3833402   (W.L. ELBAN)
(150)   US 3821013   (CELANESE)
(151)   US 3770488   (R.T. PEPPER)
(152)   US 3573961   (HAWKINS)
(153)   US 3960592   (IMPERIAL CHEMICAL INDUSTRIES)
(154)   US 4005172   (IMPERIAL CHEMICAL INDUSTRIES)
(155)   US 4008299   (IMPERIAL CHEMICAL INDUSTRIES)
(156)   GB 1320908   (ENGLISH ELECTRIC CO)
(157)   JA 7471218   (MITSUBISHI RAYON)

(158)   JA 7589695   (TOHO BESLON)
(159)   US 3806489   (RHONE-PROGIL)
(160)   US 3957716   (HERCULES)
(161)   US 4216262   (GREAT LAKES CARBON)
(162)   GB 2005237   (ROLLS-ROYCE)
(163)   FR 2203777   (BAYER)
(164)   US 3501491   (J.M. ROSS)
(165)   US 3695916   (UNITED AIRCRAFT CORP.)
(166)   US 3762941   (CELANESE)
(167)   US 3853600   (CELANESE)
(168)   US 3837904   (GREAT LAKES CARBON)
(169)   US 3844822   (CELANESE)
(170)   FR 211022   (TRW)
(171)   EP 15729   (CELANESE)
(172)   DE 2537272   (VOGEL F.)
(173)   FR 2231414   (MITSUBISHI RAYON)
(174)   DE 2205122   (KUREHA KKKK)
(175)   FR 2231414   (MITSUBISHI RAYON)
(176)   GB 1301101   (SECRETARY OF STATE FOR DEFENSE)
(177)   CH 514500   (LONZA-WERKE)

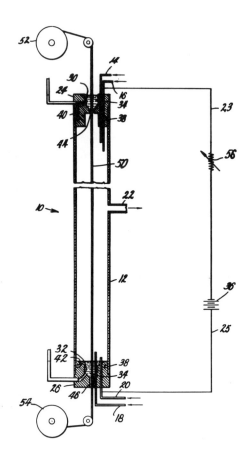

Apparatus provided with mercury seals for
the preparation of Boron fibres by CVD

CHAPTER 3

Boron Fibres

1. Summary of the Prior Art

The first attempts were aimed at the preparation of pure boron fibres but it was soon realized that due to the brittleness of the material, continuous filaments of pure boron had no practical significance; only relatively short-lengthed boron whiskers have been prepared on a semi-industrial scale and have found some technological application.
All methods known up to date for the preparation of continuous boron filaments call for a vapor deposition method of amorphous boron on a heated core, resulting in a composite fibre material. The oldest patent publication dates from the late twenties (1) and deals with pyrolysis of a boron halide on a heated tungsten wire. But the high temperature (1500°C) necessary for the pyrolysis gave rise to surface reaction between the core material and the boron with the formation of a brittle WB intermediate layer.
Furthermore the deposited boron showed a pronounced crystalline character resulting in poor mechanical properties of the filaments.

Important progress was obtained in the USA by the introduction of the chemical vapor deposition (CVD) method for boron, which could be effected at lower temperatures thus avoiding the said drawbacks of surface reaction and crystallisation (2).
A further improvement consisted in the provision of intermediate layers of boride between the W core and the boron coating to eliminate all surface reaction (3)(4). Notwithstanding these improvements the WB composite filaments produced thus far suffered from severe shortcomings embracing only mediocre mechanical properties, such as

- irregular deposits of boron
- spontaneous crackup of the boron coating due to high residual stresses in the W core.

2. Developments since 1970

The CVD method has now become the basic way for the preparation of boron filaments and up to now no fundamental modifications in the process have been reported. Most efforts for improving boron filamentary products have been done in the USA and were focused on the following aspects, which will be discussed in detail hereafter.

1° use of the other core materials
2° improvements of the CVD process and of the related apparatus
3° aftertreatments

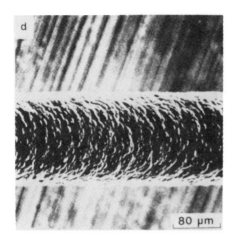 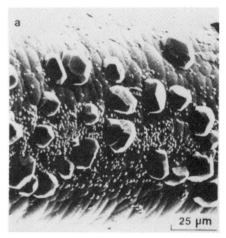

Surface morphology of boron fibres obtained by CVD

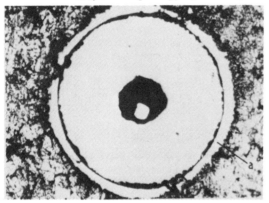

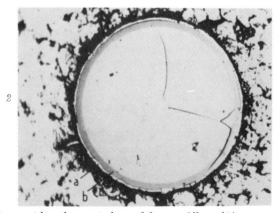

Reaction zone between titanium matrix and boron fibre (1),
or B_4C-coated boron fibre (2), after annealing 50 h at
845°C. (a) Reaction zone, (b) diffusion barrier.

2.1. Use of other core materials.   Substitution of the relatively heavy and expensive W core by more appropriate materials was one of the major problems to be solved in order to render boron filaments commercially more attractive. Several suggestions in that direction were made such as the use of carbon-or tungsten coated silica fibres (5) (6) or beryllium wire (7). But the major breakthrough came with the appearance of commercially available carbon fibres, which possessed the desirable characteristics of electrical conductivity, hot strength and low density (8). It was however observed that during the boron deposition serious degradation of the carbon filaments occured resulting in non uniform coatings with stringlike appearance. Said phenomenon was caused by fractures in the carbon core, enhanced by expansion of the initial B coating, thus burning "hot spots" along the filaments.
Most techniques developed to reduce this phenomenon implied a precoating of the carbon filaments, e.g. with pyrolytic graphite (9) (10) (11) (12) or metal carbides (13) (14); in another embodiment it was proposed to use glassy carbon filaments possessing better high temperature expansion properties than the normal carbon fibers (15).

2.2. Improvements in the CVD process and related apparatus.   These were primarily concerned with special heating techniques of the filamentary core to obtain more uniform coatings, such as

- radio frequency induction heating of selected sections of the cores to provide controlled temperature profiles (16)
- combination of resisting heating of the core with supplementary external heating (17) (18) or high frequency heating (25).

Other improvements were related to different process parameters e.g. pretreatment of the core material by heating in a nitrogen atmosphere (19); increasing the deposition rate by addition of catalytic amounts of WF_6 to the gaseous atmosphere (20); or to constructional details of the apparatus e.g.

- the provision of liquid seals at the cell ends to avoid any friction between filament and sealing devices (21)
- the provision of cooling stages (22)
- development of multifilament apparatus (23) or multisection apparatus (24)

2.3. Post-treatments of boron filaments.   Two major types of post-treatments can be distinguished

1° chemical and/or thermal post-treatments
2° post-treatments involving supplementary coating of the boron filaments

1° The chemical post-treatments were intended to remove surface defects (flaws) which affected the fracture properties of the filaments, and consisted of an etching or polishing action by some chemicals (27) (31) (36); while the thermal post-treatments were aimed at the elimination of residual stresses (26).

2° The provision of supplementary coatings on the other hand became necessary to protect the boron layer against reaction at higher temperature with matrix material, especially metal matrices. These coatings consist of diffusion barrier layers composed of Al_2O_3 (28) or of different carbides, borides or nitrides (31) (32) (33) (37) and were obtained by CVD.

3. Properties and Uses

The properties of composite boron filaments can be summarised as follows
- high strength to weight ratio,
- high melting temperature,
- high modules of elasticity,
- retention of strength up to temperature of 1000°C.

Boron filaments are mainly used for the manufacture of inorganic composite materials.
Boron fibre reinforced organic matrices are well known (2) but the use of such materials is limited by the properties of the resins at high temperatures. Therefore the specific high temperature properties of the boron fibres could only be fully exploited by combining them with high melting matrix material for the production of exotic composites.
In the early seventies attempts were made to produce boron-light metal composite structures for aerospatial applications. But the chemical interaction at higher temperature between the boron and matrix metal which tends to decrease the strength of the fibres and to weaken the cohesion of the material, forbade the use of uncoated fibres. It was therefore proposed to provide the boron filaments with diffusion barrier layers as discussed in 2.3. But this in turn could negatively affect strength properties of the boron fibre itself (29). Investigations are still going on to determine optimal conditions of coating thickness, composition, etc.

4. Latest Developments and Trends

During the period from 1980 up to now, no further substantial developments in the boron fibre manufacture have been reported. One publication describes the production of pure boron fibres by CVD of boron on a tungsten core followed by longitudinal splitting and chemical dissolution of the W core (30).

Most of the recently published literature however deals with investigations concerning the chemical, physical or mechanical behaviour of the boron filament when incorporated in the matrix, or with the strength properties of the composites (34) (35), but this will be discussed in more detail in the relevant chapters pertaining to the composite materials.

As can be deduced from the different publications cited in reference most of the pioneering work in the field of boron fibre manufacture has been done in the USA (UNITED AIRCRAFT CORP.; AVCO; GENERAL ELECTRIC; TEXACO; NORTHROP a.o.). More recently some activity was also reported in FRANCE (SNECMA; SOC. NAT. DES POUDRES ET EXPLOSIFS) and to a lesser degree in FRG.

As far as can be concluded from the present literature study, little or no substantial work has been done in Japan.

*   *

*

REFERENCES TO CHAPTER 3

(1)	NL 19624	(PHILIPS) = FR 617443
(2)	GB 1051883	(TEXACO)
(3)	US 3451840	(R.I. HOUGH)
(4)	GB 1108659	(W.M. WEIL)
(5)	US 3620836	(GENERAL ELECTRIC)
(6)	US 3787236	(UNITED AIRCRAFT)
(7)	US 3741797	(GENERAL TECHNOLOGY CORP)
(8)	DE 1954480	(UNITED AIRCRAFT)
(9)	US 3679475	(UNITED AIRCRAFT)
(10)	US 4142008	(AVCO CORP)
(11)	US 4163583	(AVCO CORP)
(12)	US 4045597	(AVCO CORP)
(13)	US 3903347	(UNITED AIRCRAFT)
(14)	US 3903323	(UNITED AIRCRAFT)
(15)	US 3811927	(GREAT LAKES CARBON CORP)
(16)	US 3572286	(TEXACO)
(17)	FR 2243907	(SOC. NAT. DES POUDRES ET EXPLOSIFS)
(18)	DE 1696101	(CONS. FUR ELECTROCHEM. IND)
(19)	NL 7100743	(AVCO)
(20)	US 3811930	(UNITED AIRCRAFT)
(21)	GB 1177854	(UNITED AIRCRAFT)
(22)	US 4031851	(J. CAMAHORT)
(23)	US 3887722	(UNITED AIRCRAFT)
(24)	FR 2029371	(ETAT FRANCAIS)
(25)	EP 18260	(SOC. NAT. DES POUDRES ET EXPLOSIFS)
(26)	NL 7507558	(NORTHROP CORP)
(27)	FR 2131858	(SNECMA)
(28)	GB 1215800	(U.S. COMPOSITES CORP)
(29)	Chem. Abstr. vol 92 ref. 10216 p	
(30)	Chem. Abstr. vol 94 ref. 161358 d	
(31)	Journal of Less common Metals 47 (1976) 207-213	
(32)	Journal of Less Common Metals 47 (1976) 215-220	
(33)	Journal of Less Common Metals 47 (1976) 221-223	
(34)	NASA Report NASA-TM-92806	
(35)	NASA Report NASA-TM-82559	
(36)	US 3698970	(UNITED AIRCRAFT CORP)
(37)	US 3917783	(NORTHROP CORP)

CHAPTER 4

Polycrystalline Refractory Oxide Fibres

Polycrystalline refractory oxide fibres, especially continuous fibres *i.e.* fibres having an infinite length versus their diameter, are very desirable owing to their combination of high theoretical tensile strength and modulus of elasticity with chemical inertness at ambient temperature and the retention of a substantial proportion of such properties after exposure to high temperatures. Many efforts have been made to produce such fibres because of an increasing demand for high-temperature structural materials, particularly as reinforcing fibres in aerospace technology, for high-temperature insulation, filters, etc.

1. Summary of the Prior Art

The strength of a polycrystalline refractory fibre is highly dependent on its microstructure (1). Pores, defects, as well as crystallite size markedly affect the microstructure and hence the properties. Since microstructure and fibre strength are strongly influenced by the method used to produce the polycrystalline fibres, suitable manufacture processes had first to be developed. The most interesting processes developed in the 1950-1970 period are briefly reviewed here because of their importance in the further development of production technology after 1970.

1.1. Manufacture processes.

1.1.1. Molten oxide process :   A process in which continuous or discontinuous oxide fibres are formed from a melt e.g. by extrusion; processes have been developed by Rolls-Royce (2,3) and Feldmühle AG (4,5).

1.1.2. Extrusion process :   A process in which a plasticized mass containing a dispersion of finely divided oxide particles, is forced through a die to form continuous fibres; the process is described by H.I. Thompson Fibre Glass Co (6, see p. 49), and National Beryllia Corp (7) and reviewed by J.E. Bailey et al. (8), particularly for producing Al_2O_3 fibres.

1.1.3. Precursor processes :   Processes in which a solution or suspension of a metal compound, e.g. inorganic and organic salts, is formed into fibres in the presence or not of an organic material; by burning, the metal compound is converted to the corresponding metal oxide and the organic material, if present, disappears.

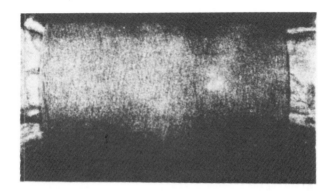

cm   0 1 2 3 4 5 6

Continuous alumina filament yarns wound
onto a collapsible, refractory bobbin

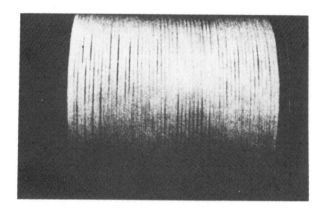

cm ˙0 1 2 3 4 5 6

Continuous alumina filament yarns after unwinding
and firing between 1000°C and 1800°C

The following processes are used :

- evaporation of colloïdal suspensions : fibres are formed by spreading a thin film of a colloïdal suspension on a flat surface which is rapidly heated; the method was developed by Horizons Inc (9-14) and Norton Co (15)

- "relic" process : organic fibres or fabrics are impregnated with metal oxide precursors; the method was developed by Union Carbide Corp (16,17)

- the Rayon Spinnerette process : FMC Corp (18-23) developed this method for producing continuous fibres in which a cellulosic material and a glass-forming oxide are extruded into a spinning bath

- fibrizing viscous fluids of metal oxide precursors by usual fibre forming methods e.g. spinning, extrusion and drawing; this approach, the most promising to obtain high-strength continuous fibres, was introduced by Horizons Inc (24-26), Hitco (27,28), Thompson Fibre Glass Corp (29) and the Babcock & Wilcox Co (30,31); subsequently Celanese Corp (32) and UKAEA (33) proposed the addition of organic polymers to the metal oxide precursor solution or dispersion.

1.2. Composition, properties and applications.   For a more or less complete survey of composition, properties and applications of refractory oxide fibres covering the 1950-1970 period, one should refer to the work of H.W. Rauch *et al.* (34), the review of J.E. Bailey *et al.* (8) and the references cited in section 1.1.

2. Developments since 1970.

Increased research into composites and the need for high temperature insulation in industrial furnaces to face the energy crisis resulted in an increased interest by industrial organisations to develop suitable ceramic fibres. Based on the successful results obtained by application of the manufacture processes mentioned above, emphasis has been put on the further development of these processes together with the development of new compositions in order to make ceramic fibres suitable for specific applications.

At the same time new applications have been found and developed. Imperial Chemical Industries, Farbenfabriken Bayer AG, Minnesota Mining & Manufacturing Company, E.I. Du Pont de Nemours, Universal Oil, Union Carbide Corp., Aluminium Company of America and recently Monsanto Company with their development of hollow fibres, are the leading organisations in this area.

Up to now, only a few Japanese organisations are dealing with ceramic oxide fibre production development.

2.1. Further development of manufacture processes.   Since 1970 very few really new production methods have been introduced, for example :

- R.W. Jech et al. (35) describe the preparation of ceramic fibres by hot extrusion of ceramic oxide particles in a tungsten matrix; by re-extrusion, length to diameter ratios of ZrO_2 fibres were as high as 640.

- W. Dannöhl (36) describes the preparation of magnesium oxide skeletons by selective oxidation of magnesium containing alloys; the skeletons are used as fibre reinforcement in composites.

- T. Shimizu et al. (37) describes the hydrothermal synthesis of potassium titanate fibres from K_2O-TiO_2 glass.

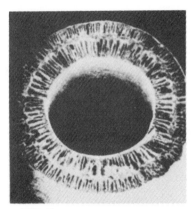 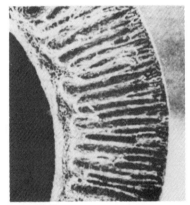

100.0 u ⊢————⏋
16-1 20.0 23 166 302

0 10.0 u ⊢—⏋
05-2 20.0 09 166 101

*Hollow fibres having radially anisotropic internal
void volume wall structures (MONSANTO CO)*

- A more promising method is reported by A.C.D. Chaklader *et al.* (38) : the report summarises attemps to produce alumina fibres from fine continuous aluminium wire by anodizing; the strength of the fibres obtained is quite close to the values reported for Saffil alumina fibres produced by I.C.I. (see section 2.1.4.).

As already mentioned, continuous ceramic fibres are most desirable. Manufacturers made useful processes available one at a time to produce fibres suitable for the applications for which they were intended. As described in section 1.1. suitable methods were : the molten oxide process, the extrusion process, the "relic" process and the process of fibrizing viscous fluids of metal oxide precursors.

Consequently available data on refractory oxide fibres are reviewed hereafter according to their method of fabrication since these methods cover almost completely the most important developments from 1970 on.

2.1.1. Molten oxide process :   The method is used only in special cases since special precautions have to be taken making thus the process very expensive.

According to G. Sporleder (5), the average strength for 4-8µ diameter Al_2O_3 fibres, extruded from the melt, is about 160 kg/mm^2.

Bjorksten Research Labs (39) describes the process of drawing melt spun fibres from pure and eutectic oxides to still finer fibres.

Y. Fujiki (40) produces crystalline potassium titanate fibres from a molten product which can be spun by the glass fibre method.

However by degrading molten alumina with silica, known glass fibre forming technology may be used; manufacturers dealing with these fibrous products are for example The Carborundum Co (41,42), Société Européenne des Produits Réfractaires (43), Johns-Manville (44,45) and The Babcock & Wilcox Co (46); the products obtained are mainly used for insulation purposes and can be used at temperatures up to about 1300°C; therefore they are only briefly mentioned herein.

2.1.2. Extrusion process :   Du Pont de Nemours (47-51) produces continuous alumina fibres for reinforcement purposes by extruding an aqueous mix containing selected alumina particles and water-soluble precursors of alumina in selected amounts.

ALCOA (52) describes the production of alumina fibres, mainly for high-temperature insulation, by extruding an alumina slip of very high solids content; in a later development (53) an organic polymer was added to the slip to produce continuous fibres.

United States Department of Energy (54) uses a slurry of metal oxide powder and a polymeric solution for extrusion; suitable apparatus is described; the fibres are used to produce molten carbonate fuel cell tiles.

In a recent development the extrusion process has been found to be very suitable for preparing hollow ceramic fibres for special purposes e.g. ultrafiltration and fluid separation devices where a certain porosity is desirable.

Monsanto (55) describes the extrusion of a sinterable inorganic material

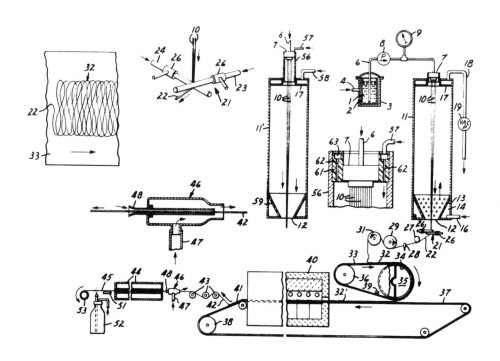

Flow chart of continuous refractory oxide fibre production
(MINNESOTA MINING AND MANUFACTURING CO)

containing polymer solution through a hollow fibre spinneret directly into
a coagulating bath; alumina, beta-alumina, cermets, etc. may be used as
inorganic material; inorganic hollow fibres having a radially anisotropic
internal void volume wall structure may be obtained (56) particularly for
applications such as fluid separations, filtrations, gassparging, fuel cells
and batteries, as well as coiled hollow fibres (57), particularly to be used
as solid electrolyte in batteries.

Du Pont de Nemours (58) produces hollow ceramic filaments for separation
devices by extrusion of a metal oxide containing polymer solution through a
die together with a combustible core such as cotton thread which is afterwards burned; rigid, porous, inorganic hollow filaments comprising about
5-50% cordierite and about 95-50% a alumina are thus produced (59).

2.1.3. "Relic" process :   Union Carbide uses the process for the preparation
of metal oxide fabrics (60), e.g. cubic zirconia and thoria fabrics, and
stabilized tetragonal zirconia fibres and textiles (61); the products are
useful as thermal insulators, battery spacers and the like.

R.B. Diver *et al.* (62) describes the fabrication of thoria effusion membranes
by this technique.

A method of making high tensile strength microcrystalline fibres and for
controlling their diameter is claimed by Versar Inc. (63).

2.1.4. Fibrizing viscous fluids of metal oxide precursors :   A great deal
of work related to the development of high strength ceramic oxide fibres
made by this process is being carried out. Indeed the process enables the
production of continuous dense fibres with few pores and consequently with
improved tensile strength and modulus.

Imperial Chemical Industries (64-66) uses a fibrizing composition having a
viscosity of at least 1 poise which comprises a solvent, a metal salt and an
organic polymer both dissolved in said solvent to produce refractory oxide
fibres, particularly Al_2O_3 and ZrO_2 fibres, for catalyst and insulation
purposes (67); additional improvements with relation to the extrusion
conditions are made (68-70); a zirconia fibre comprising alumina and yttria
has also been prepared (71); the addition of an organic silicone to the
above mentioned fibrizing composition is claimed for the preparation of
refractory fibres containing metal oxide and silica (72); in a further
improvement an aqueous solution of reduced anion content has been used (73);
in another process refractory spinel fibres are prepared using a viscous
composition which comprises an organic solution of spinel forming metal
oxide precursors (74); in another development anisometric particles are
added to the fibre precursor composition to orientate the particles during
formation of the fibre by the extrusion or drawing of the liquid phase (75).

Initially Bayer AG used a spinning solution of metal compounds to prepare
ceramic filaments (76); later on a more suitable fibrizing composition was
developed comprising a solution or sol containing one or more metal compounds
and a linear polymeric fibre-forming material having a degree of polymerisa-
tion of at least about 2000 (77); the process is used to produce reinforcement fibres comprising a metal oxide and a finely divided disperse phase
distributed throughout the oxide phase (78), to produce silica fibres (79),
to produce alumina fibres containing a proportion of silica which are
particularly useful as reinforcement in a plastic, metal, glass or ceramic
matrix (80) and to produce quartz glass or cristobalite fibres for high-
temperature-resistant structures reinforced or insulated therewith (81, 82);

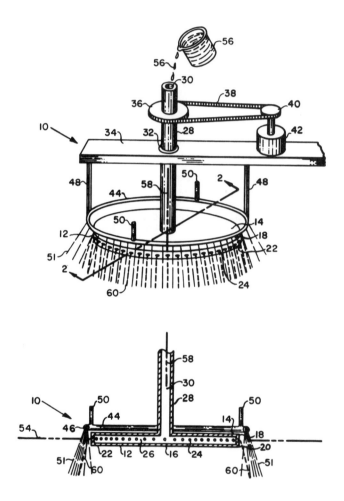

Spinning device for manufacturing ceramic oxide fibres from a solvent solution (KENNECOTT CORP)

in a further improvement an organic solution is applied after spinning and
before sintering in order to increase the tensile strength of the resulting
alumina fibres (83).

Minnesota Mining and Manufacturing prepares transparent refractory fibres
with specific compositions which are useful to form fabrics or as reinforcement for composites; the fibres are made by shaping and dehydratively
gelling, particularly by extruding in air, viscous aqueous metal compound
solutions or sols, and heating the resulting gelled fibres in a controlled
manner; to obtain continuous high strength refractory fibres the drawn green
fibres are accumulated in a released, loose, unrestrained, random or orderly
configuration on a moving substrate which afterwards passes into a firing
furnace (85); refractory fibres of zirconia and silica mixtures (86),
aluminium borate or borosilicate fibres (87), high density thoria-silica-
metal (III) oxide fibres (88), non-frangible alumina-silica fibres (89),
alumina-chromia-metal(IV)oxide fibres (90) and titanium dioxide fibres
(91) are produced according to the above mentioned process.

Universal Oil Products prepares flexible, porous alumina fibres of high
surface area which are employed as a catalyst or a catalyst support; one
method uses a fibre drawing composition consisting of an alumina sol and a
soluble fibre-forming organic polymer (92,93); in another method a mixture
of an alumina sol and hexamethylenetetramine is used (94, 95); the resulting
green fibres are calcined in air at a temperature of from about 300°C to
about 1000°C; an apparatus for producing such high surface area fibres has
been described (96).

In addition to the most important developments described above, several
other manufacturers have proposed interesting improvements to prepare
refractory oxide fibres from viscous salt solutions : The Carborundum Co
incorporates lactic acid in the said solutions in order to increase the
metal salt content thereof without affecting viscosity (97);

subsequently Kennecott Corporation developed a process and apparatus for
manufacturing fibres having an average diameter of less than 5 microns from
such fiberizable solutions (98);

another approach is proposed by United Aircraft Corporation for the preparation of alumina filaments in which the viscous salt solution is formed
into filaments by extrusion directly into a coagulating bath (99);

Sumitomo Chemical Company produces alumina and alumina-silica fibres for
reinforcement purposes and having excellent mechanical strength and heat
resistance by spinning an organic solution of polyaluminoxanes or of a
mixture of polyaluminoxanes and one or more kinds of silicon-containing
compounds (100, 101);

another Japanese development is related to the preparation of spinnable
viscous solutions of metal alcoholates by hydrolysis and TiO_2 - SiO_2,
Al_2O_3 - SiO_2 and ZrO_2 - SiO_2 refractory fibres are thus produced (102-105);

in conclusion aluminium phosphate fibres are prepared from viscous solutions
as described by J.D. Birchalli et $al.$ (106) and others (107) and silicon
dioxide fibres are produced from aqueous alcali silicate solution by AKZO (108).

Further processes to prepare ceramic oxide fibres are related to the
preparation of gel fibres; extrusion of partially gelled sols as proposed

by P.F. Becher *et al.* (109), spray drying of viscous aqueous sols as described by the UKAEA (110) and the immersion of a gel of aluminium hydroxide fibrilles in an organic polymer solution capable of floculating the gel with orientation of the fibrilles, continuous alumina filaments being obtained by drawing the gel through the said polymer solution as described by the Société de Fabrication d'Eléments Catalytiques (111).

2.2. Composition, properties and uses.   Polycrystalline oxide fibres are mainly used for catalyst, insulation and reinforcement purposes. Properties and uses of alumina fibres have been for example reviewed by J.F. Bascon *et al.* (38) and W.R. Symes (112). A general description of "Saffil" alumina and zirconia fibres developed by Imperial Chemical Industries (see section 2.1.4.) has been given in a periodical (113). A review of ceramic alumina-silica fibres containing 45 to 95% alumina has been published by A. Eschner (114).

However for complete information and for a general survey of compositions, properties and uses of polycrystalline oxide fibres produced up to now, one should refer to section 2.1. and the references cited therein.

In addition it is indicated here that special fibre treatments are developed for improving some properties of ceramic fibres as examplified hereafter : du Pont de Nemours applies a vitrified coating to prepare high-strength fibres especially suitable for reinforcement uses (115);

Union Carbide Corp describes a process for the preparation of zircon coated zirconia fibres having improved high-temperature properties and are ideally suited as thermal insulation (116);

The Carborundum Company produces chromium oxide coated fibres to improve heat resistance and shrink resistance, while, for the same purposes, Agency of Industrial Sciences and Technology describes a Japanese development consisting of treating ceramic fibres with phosphorus compounds (117) or with lithium and/or magnesium containing solutions (118);

Bjorksten Research Labs applies a thermal treatment to increase the modulus of elasticy of ceramic fibres consisting of Al_2O_3 and CaO (119).

2.3. Developments in Japan.   Besides the few Japanese work already indicated in the foregoing sections, additional developments have recently been described by Denki Kagaku Kogyo K.K. for manufacturing alumina fibres (120), alumina or zirconia fibres (121) and alumina-silica fibres (122), by Nippon Asbestos Co for manufacturing heat-insulating alumina fibres (123) and by Asahi Glass Co for manufacturing calcium oxide-chromium oxide fibres suitable for reinforcing plastics and ceramics (124).

As already mentioned above it can be seen that up to now, Japanese organisations are only involved to a small extent in the development of polycrystalline oxide fibres, most work being done in Europe and the USA.

*   *

*

REFERENCES TO CHAPTER 4

(1) W.L. LACHMAN and J.P. STERRY, Chemical Eng. progr., 58 (10)
     37-41 (1962)
(2) GB 1069472   (ROLLS-ROYCE LTD)
(3) GB 1078742   (ROLLS-ROYCE LTD)
(4) DE 1671891   (FELDMUHLE AG)
(5) G. SPORLEDER, Chemical Abstr. 78 (13) 163.251g (1973)
(6) L.R. McCREIGHT, H.W. RAUCH and W.H. SUTTON, Ceramic and graphite
     fibres and whiskers, Academic Press, New York, 1965
(7) GB 1001003   (NATIONAL BERYLLIA CORP)
(8) J.E. BAILEY and H.A. BARKER, Chem. Br. 10 (12) 465-70 (1974)
(9) US 3096144   (HORIZONS INC.)
(10) US 3082103   (HOTIZONS INC.)
(11) US 3082099   (HORIZONS INC.)
(12) US 3082055   (HORIZONS INC.)
(13) US 3082054   (HORIZONS INC.)
(14) US 3082051   (HORIZONS INC.)
(15) FR 1274807   (NORTON COMPANY)
(16) US 3385915   (UNION CARBIDE CORP)
(17) GB 1144033   (UNION CARBIDE CORP)
(18) Anon., Chemical & Eng. News, 44 (1) 32 (1966)
(19) US 3565749   (FMC CORP)
(20) US 3428719   (FMC CORP)
(21) GB 1173740   (FMC CORP)
(22) GB 1155292   (FMC CORP)
(23) GB 1064271   (FMC CORP)
(24) US 3180741   (HORIZONS INC.)
(25) US 3271173   (HORIZONS INC.)
(26) US 3311689   (HORIZONS INC.)
(27) US 3311481   (HITCO)
(28) US 3416953   (HITCO)
(29) GB 1030232   (THOMPSON FIBRE GLASS CORP)
(30) US 3322865   (THE BABCOCK & WILCOX CO)
(31) US 3503765   (THE BABCOCK & WILCOX CO)
(32) FR 2002846   (CELANESE CORP)
(33) FR 1578319   (UKAEA)
(34) H.W. RAUCH, W.H. SUTTON and L.R. McCREIGHT, Ceramic fibres and
     fibrous composite materials, Academic Press, New York, 1968
(35) R.W. JECH, J.W. WEETON and R.A. SIGNORELLI, Amer. Cer. Soc. Bull.,
     49 (11) 983-87 (1970)
(36) CH 588416   (W. DANNÖHL)
(37) T. SHIMIZU, K. HASHIMITO and H. YANAGIDA, Cer. Abstr., 55 (5/6)
     107f (1976)
(38) A.C.D. CHAKLADER, H.H. HAWTHORNE and S.K. NHATTACHARYA, Composites,
     12 (4) 272-74 (1981)
(39) US 4104355   (BJORKSTEN RESEARCH LABORATORIES)

45

(40)   US 4265872   (Y. FUJIKI)
(41)   EP 7485   (THE CARBORUNDUM CO)
(42)   FR 2182864   (THE CARBORUNDUM CO)
(43)   FR 2481263   (SOCIETE EUROPEENNE DES PRODUITS REFRACTAIRES)
(44)   FR 2450795   (JOHNS-MANVILLE CORP)
(45)   FR 2348899   (JOHNS-MANVILLE CORP)
(46)   FR 2272968   (THE BABCOCK & WILCOX CO)
(47)   FR 2026821   (E.I. DU PONT DE NEMOURS)
(48)   US 3853688   (E.I. DU PONT DE NEMOURS)
(49)   US 3808015   (E.I. DU PONT DE NEMOURS)
(50)   FR 2237841   (E.I. DU PONT DE NEMOURS)
(51)   US 3953561   (E.I. DU PONT DE NEMOURS)
(52)   US 3705223   (ALUMINUM COMPANY OF AMERICA)
(53)   US 4071594   (ALUMINUM COMPANY OF AMERICA)
(54)   GB 2055356   (UNITED STATES DEPARTMENT OF ENERGY)
(55)   US 4222977   (MONSANTO CO)
(56)   US 4175153   (MONSANTO CO)
(57)   EP 47640   (MONSANTO CO)
(58)   FR 2381000   (E.I. DU PONT DE NEMOURS)
(59)   WO 81/00523   (E.I. DU PONT DE NEMOURS)
(60)   US 3663182   (UNION CARBIDE CORP)
(61)   US 3860529   (UNION CARBIDE CORP)
(62)   R.B. DRIVER and E.A. FLETCHER, Amer. Cer. Soc. Bull. 56 (11)
       1019-20 (1977)
(63)   US 4104395   (VERSAR INC)
(64)   GB 1360197   (IMPERIAL CHEMICAL INDUSTRIES)
(65)   GB 1360199   (IMPERIAL CHEMICAL INDUSTRIES)
(66)   GB 1360200   (IMPERIAL CHEMICAL INDUSTRIES)
(67)   FR 2213253   (IMPERIAL CHEMICAL INDUSTRIES)
(68)   FR 2176041   (IMPERIAL CHEMICAL INDUSTRIES)
(69)   US 3992498   (IMPERIAL CHEMICAL INDUSTRIES)
(70)   GB 1470292   (IMPERIAL CHEMICAL INDUSTRIES)
(71)   GB 1360198   (IMPERIAL CHEMICAL INDUSTRIES)
(72)   US 4094690   (IMPERIAL CHEMICAL INDUSTRIES)
(73)   GB 2059933   (IMPERIAL CHEMICAL INDUSTRIES)
(74)   FR 2144760   (IMPERIAL CHEMICAL INDUSTRIES)
(75)   GB 1354884   (IMPERIAL CHEMICAL INDUSTRIES)
(76)   FR 2039169   (BAYER AG)
(77)   US 3846527   (BAYER AG)
(78)   US 4010233   (BAYER AG)
(79)   FR 2103411   (BAYER AG)
(80)   US 3982955   (BAYER AG)
(81)   US 4104045   (BAYER AG)
(82)   US 4180409   (BAYER AG)
(83)   FR 2229788   (BAYER AG)
(84)   D.D. JOHNSON, Chem. Abstr. 83 (20) 234-167.8775 (1975)
(85)   US 3760049   (MINNESOTA MINING AND MANUFACTURING CO)
(86)   US 3793041   (MINNESOTA MINING AND MANUFACTURING CO)
(87)   US 3795524   (MINNESOTA MINING AND MANUFACTURING CO)
(88)   US 3909278   (MINNESOTA MINING AND MANUFACTURING CO)
(89)   US 4047965   (MINNESOTA MINING AND MANUFACTURING CO)
(90)   US 4125406   (MINNESOTA MINING AND MANUFACTURING CO)
(91)   US 4166147   (MINNESOTA MINING AND MANUFACTURING CO)
(92)   US 4250131   (UOP INC)
(93)   US 3652749   (UNIVERSAL OIL PRODUCTS CO)
(94)   US 3632709   (UNIVERSAL OIL PRODUCTS CO)
(95)   US 3814782   (UNIVERSAL OIL PRODUCTS CO)
(96)   US 3614809   (UNIVERSAL OIL PRODUCTS CO)

(97)   FR 2359085   (THE CARBORUNDUM CO)
(98)   EP 31656   (KENNECOTT CORPORATION)
(99)   US 3865917   (UNITED AIRCRAFT CORPORATION)
(100)   US 4101615   (SUMITOMO CHEMICAL COMPANY)
(101)   US 4152149   (SUMITOMO CHEMICAL COMPANY)
(102)   S. SAKKA and M. TASHIRO, Chem. Abstr., 86 (16) 33.344u (1977)
(103)   K. KAMIYA and S. SAKKA, Chem. Abstr., 87 (18) 139.988t (1977)
(104)   K. KAMIYA and S. SAKKA, Chem. Abstr., 88 (16) 109.413e (1978)
(105)   JP 7747052   (M. SAKUHANA et al.)
(106)   US 4008299   (IMPERIAL CHEMICAL INDUSTRIES) (ICI)
(107)   NASA, NASA TECH BRIEF, B 74-10.185 (1974)
(108)   FR 2446334   (AKZO N.V.)
(109)   P.F. BECHER, J.H. SEMMERS, B.A. BENDER and B.A. MACFARLENE,
        Mater.Sc. Res., (11) 79-86 (1978)
(110)   US 3704147   (UKAEA)
(111)   FR 2088130   (SOCIETE DE FABRICATION D'ELEMENTS CATALYTIQUES)
(112)   W.R. SYMES, GAS WÄRME INT., 30 (7-8) 371-78 (1981)
(113)   ANON., SPRECHSAAL, 107 (13) 604-07 (1974)
(114)   A. ESCHNER, GAS WÄRME INT., 30 (7-8) 357-62 (1981)
(115)   US 3849181   (E.I. DU PONT DE NEMOURS CO)
(116)   US 3861947   (UNION CARBIDE CORPORATION)
(117)   DE 2808373   (AGENCY OF INDUSTRIAL SCIENCE & TECHNOLOGY)
(118)   JP 77114727   (AGENCY OF INDUSTRIAL SCIENCE & TECHNOLOGY)
(119)   JP 8023007   (BJORKSTEN RESEARCH LABORATORIES)
(120)   JP 8045809   (DENKI KAGAKU KOGYO K.K.)
(121)   JP 8045808   (DENKI KAGAKU KOGYO K.K.)
(122)   JP 8020239   (DENKI KAGAKY KOGYO K.K.)
(123)   JP 8109427   (NIPPON ASBESTOS CO)
(124)   JP 7527479   (ASAHI GLASS CO)

CHAPTER 5

Polycrystalline Refractory Carbide, Nitride and Boride Fibres

Based on the increasing demands for high-temperature structural materials continuous carbide, nitride and boride fibres or filaments have been developed especially as reinforcing fibres for metal composite materials. This development which started in the sixties and still continues, attempts to form filaments with better properties, emphasis being laid on the production of continuous silicon carbide filaments because of their structural and oxidation resistant properties.

1. Summary of the Prior Art

As difficulties were met by using boron filaments for the preparation of fibre reinforced metal composite materials (see chapter 3 : Boron fibres), investigations were directed to prepare filaments of material with structural properties similar to those of boron. Materials of particular interest were the carbides, especially silicon carbide and boron carbide.
The preparation of these refractory fibres by conventional fibre-forming techniques were very difficult owing to the high melting point and low sinterability and other processing routes had to be found. Initially the following three methods were developed and made useful to produce the refractory filaments involved.

1.1. Chemical vapor deposition (C.V.D.) :   According to this process a heated core in filament form e.g. a tungsten wire is continuously drawn through a sealed deposition chamber wherein a vaporized reactant is introduced, which is decomposed and deposited on the core to produce a composite filament; processes and apparatus are developed mainly for the deposition of silicon carbide on a core using alkyl chlorosilanes or mixtures of chlorosilanes with hydrocarbons as vaporized reactants (1-10). Similarly boron carbide filaments are prepared using mixtures of boron trichloride and methane or carboranes as vaporized reactions (11, 12). A modified method to prepare refractory carbide filaments by C.V.D. was introduced (13, 14) : in the process no core filament is used; the decomposition products are deposited on a rotating mandrel having helical grooves formed thereon and are continuously peeled from said rotating mandrel in the form of a continuous filament of controlled thickness, uniformity and purity.

1.2. Chemical conversion of a precursor fibre in the presence of a reacting gas, liquid or solid material.   In one method refractory carbide fibres are prepared by heating carbon fibres in the presence of a carbide forming metal including silicon (15) or in the presence of vaporized compound of a carbide forming element e.g. a halide of silicon, boron or transition element (16, 17).

In another method metal nitride filaments are prepared by the nitriding of
ultra fine refractory metal wires (18); similarly boron nitride fibres are
produced by nitriding of boron oxide fibres (19); in a modification carbide
or boride filaments are obtained by contacting a molten filament of an
element with a solid or gaseous reactant of the appropriate type e.g. carbon
is reacted with a molten boron filament to produce the corresponding boron
carbide filament (20). In another method metal carbide and nitride fibres
are produced using the "relic" process (see Chapter 4, page 37) : a pre-
formed organic polymeric material is impregnated with a metal compound and
heated in a non-oxidising or nitriding atmosphere to obtain the corresponding
metal carbide or nitride fibres (21, 22).

1.3. Fibrizing spinnable precursor solutions :   Although the process was
developed chiefly to prepare refractory oxide fibres (see Chapter 4, page 37)
carbide fibres can also be prepared according to said process (23-25).

2. Developments since 1970.

In the early seventies research in relation to the available processes
indicated in section 1. was still going on and further improvements were
developed, mainly by US and French manufacturers. Emphasis was laid on the
CVD and the chemical conversion processes for the preparation of carbide,
especially silicon and boron carbide, boride and nitride filaments.
However in the mid seventies a Japanese chemist, S. YAJIMA, developed a
totally different approach to prepare strong, continuous fibres of silicon
carbide as will be described later here. Consequently a dramatic change in
production technology was introduced and Japan became the leading country
with relation to silicon carbide fibre development. Interest in the former
processes thus decreased drastically as will follow from a review given
hereafter.

2.1. Further development of the CVD process.   New approaches in the
field are related to :

1° Improved deposition conditions and related apparatus, particularly
   for the preparation of silicon carbide filaments (26-29).

2° Improved core materials (30, 31) or deposition of a thin buffer layer
   between the core material and the silicon or boron carbide deposit
   (32, 33).

3° Preparation of filaments other than silicon or boron carbide e.g.
   titanium nitride, titanium boride, aluminium boride or boron silicide
   (34, 35), boron-carbon compounds having a carbon content by weight
   between 21 and 35 percent (36) and zirconium carbide (37); recently a
   hollow boron nitride fibre is produced by depositing a coating of boron
   nitride on carbon fibre and subsequently heating the coated fibre in a
   gas which reacts with carbon to drive off the carbon (38).

4° Surface treatment of silicon carbide filaments to increase their strength
   (39) or to make them more suitable for incorporation within matrices,
   particularly within metal matrices, the latter being obtained by
   depositing thin barrier layers on the silicon carbide filaments (40, 41).

2.2. Further development of the chemical conversion process.   The process
is mainly used to prepare boron carbide (42) and boron nitride (43) filaments;
high modulus boron nitride fibres are prepared by heating partially nitrided
fibres in an inert atmosphere while subjecting them to longitudinal tension
(44); fibres consisting of boron nitride and other constituents such as

silica may also be produced (45); in a modification boron nitride filaments
are prepared by surface nitriding boron filaments on which a boron oxide
coating has previously been formed (46); in other modifications partially
nitrided boron oxide fibres are heated in an atmosphere of an amine, e.g.
methylamine, to prepare boron carbide fibres (47), or boron nitride fibres
may be further reacted with transition metal halides to form transition metal
nitride fibres (48, 49).

The process is also used for the preparation of : titanium carbide filaments
by reacting titanium filaments with carbon-containing vapor (50);silicon
carbide, silicon nitride or boron silicide filaments by contacting a silicon
filament with the corresponding gaseous reactant (51), or by heating carbon
coated silica containing fibres (52) or by immersing carbon fibres in a bath
of molten silicon (53) or by heating carbon fibres coated with silicon
powder and a fugitive resin (54) and molybdenum boride filaments by
boronization of molybdenum wires (55).
The "relic" process is used to produce flexible metal carbide fabrics, e.g.
zirconium, boron and silicon carbide fabrics (56).

2.3. Further development of the spinnable precursor process.   Fibre forming
compositions containing a metal compound and an organic polymer are developed
to produce metal carbide or nitride fibres (57) and more especially boron
carbide fibres (58). Silicon carbide fibres may be prepared by spinning a
mixture of molten pitch and silicon powder and carbonizing the resulting
fibre (59). Fibres of silicon carbide containing metallic silicon, carbon or
silicon nitride, or fibres of boron containing metallic boron, carbon or
boron nitride, or similar fibres of the metal carbides and mixtures thereof
are prepared by fibrizing a composition comprising an acrylonitrile polymer
and the metal powder or a mixture of metal powders and subsequently calcining
the resulting fibre under tension (60). Recently it has been proposed to
prepare hollow carbide, boride or nitride fibres by extruding through a
hollow spinneret a solution of a fibre forming organic polymer containing,
in dispersed form, a sinterable inorganic material (61-64).

2.4. Development of new processes.   Of the refractory fibres involved,
silicon carbide filaments are the most promising in the possible application
for reinforcement because of the high thermal stability, good mechanical
properties and excellent oxidation resistance up to 1500°C of this material.

The processes described above were however not entirely satisfactory to
produce silicon carbide filaments : the disadvantages of CVD and chemical
conversion processes are the relatively thick silicon carbide fibres obtained
which are difficult to handle, and their high manufacturing cost which
prevents their broad application; spinning complex precursor solutions
presents problems with respect to mechanical strength and purity due to poor
sinterability of the material involved. Moreover high production rates cannot
be reached. In an attempt to increase the production rate it was proposed to
produce silicon carbide fibres by continuously passing a preheated filamentary core material through a liquid organo-silicon halide, the temperature
of the core being high enough to cause the organo-silicon halide to boil,
decompose and deposit the silicon carbide coating (65); however the problems
are not entirely solved because a core is still needed and consequently
composite fibres are obtained.
Finally a real breakthrough in production technology was claimed to be found
in Japan by S. YAJIMA.
In 1975 a preliminary summary of his method was published in Japan and
subsequently an article entitled "Revolutionary SiC fibre", describing the

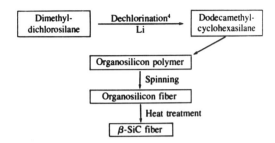

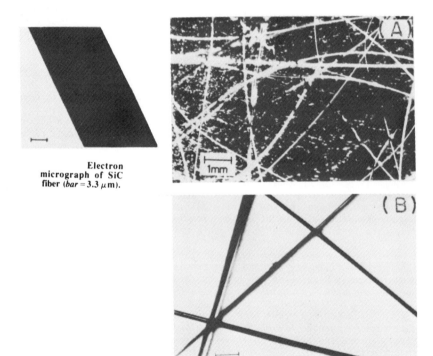

Electron micrograph of SiC fiber (*bar* = 3.3 μm).

Optical photomicrographs of (A) polycarbosilane fibres and (B) SiC fibres

Synthesis of continuous SiC fibre and related micrographs thereof according to S. YAJIMA

extraordinary features of the fibres obtained, appeared in the periodical Nature (66). According to S. YAJIMA continuous silicon carbide fibres were first obtained by heat-treating a polycarbosilane polymer which had been synthesized from dimethyldichlorosilane; the resulting fibres had an ultra-fine crystallite structure and their strength matches that of silicon carbide whiskers. In 1976 more details were published (67) and patent applications covering this method as well as the preparation of other high molecular weight organosilicon polymers from which silicon carbide fibres may be obtained, were filed (68-70). However the approach to prepare continuous silicon carbide fibres by heat-treating a carbosilane polymer precursor fibre had already been proposed and patented in 1974 in Germany by BAYER A.G. (71); the carbosilane was prepared by conversion of organo silicon compounds e.g. methyl chlorosilanes at temperatures between 400 and 1200°C.
In addition BAYER A.G. patented the preparation of continuous fibres compri-sing a mixture of silicon carbide and silicon nitride from silazane (72) and carbosilazane resins (73). Nevertheless it was only after the successful results obtained and published by S. YAJIMA that a tremendous development of the process was started in Japan and subsequently in the USA. Continuing research was especially directed to :

1° investigations for suitable silicon polymers, including improved preparations thereof, useful as pre-silicon carbide polymers, together with the development of new silicon polymers therefor (74-85)

2° preparation of continuous silicon carbide fibres having improved physical properties from polycarbosilanes modified with boron (86, 87) or transition metals e.g. polymetallocarbosilanes or copolymers com-prising a polycarbosilane portion and a polymetallosiloxane portion (88-91).

3. Post-treatments

Post-treatments have been applied to the fibres particularly to :

1° increase their strength

2° make them more suitable for incorporation within matrices, particularly metal matrices

Improvement in fibre strength may be achieved by heat treatment of the fibres under a controlled degree of tension (92, 93) or by surface treatment of the fibres comprising an electrolytic treatment (39).
In order to make the fibres more suitable for incorporation within metal matrices thin barrier layers e.g. hafnium carbide or hafnium nitride layers (40) or a carbon-rich silicon carbide layer (41), have been deposited.
It is noted here that the post-treatments described above have been applied to fibres produced by CVD or chemical conversion processes. For silicon carbide fibres produced from organsilicon precursor fibres such post-treatments have not been described up to now.

4. Properties and Uses

With reference to silicon carbide fibres which are the most important and the best developed fibres at this moment, the most striking properties are their high thermal stability, their good mechanical properties and their excellent resistance to oxidation and chemical attack. Moreover the new production route developed by S. YAJIMA made fibres with markedly improved mechanical properties available : the tensile strength and Young's modulus of the fibres obtained are 350 kg/mm^2 and 30 tons/mm^2, respectively (67); silicon carbide

fibres obtained according to the CVD process have only tensile strengths between approximately 200 and 280 kg/mm^2 (6).
In order to have a more detailed description of properties of silicon carbide fibres as well as properties of other carbide, nitride and boride fibres, one should refer to the foregoing sections and the references cited herein. The fibres reviewed in this chapter are mainly used for reinforcement, particularly to reinforce metals. They have however also been used for insulation applications. In this respect boron nitride fibres have been particularly used for the preparation of boron nitride - boron nitride composite bodies (94) which are especially useful in electrical and electronic applications requiring a material which simultaneously acts as an electrical insulator and a thermal conductor. Recently boron nitride fibre mats have been produced for use as electric cell separators in a lithium sulfide battery (95-97).

5. Latest Developments and Trends

According to R.W. RICE (98) fibres of refractory borides, carbides and nitrides or mixtures thereof can be produced by a suitable selection of one or a mixture of polymers having the ability to be converted to the corresponding borides, carbides and nitrides by heat treatment. Besides the reduction in production costs, unique microstructural control of the resulting fibres should be possible; this combined with the absence of additives for their formation should provide unique mechanical properties both at room and elevated temperatures.

Consequently further developments will be related to organometallic precursor fibre methods and more specifically to detailed investigations of different basic polymeric materials as possible refractory precursor candidates. In the near future silicon carbide fibre preparation will be further developed according to the basic principles outlined above, page 51, especially in Japan and the USA, as shown in the latest patent applications filed by UBE INDUSTRIES LTD (99) and DOW CORNING CORP (100) to name the most important manufacturers at the moment.

*   *

*

REFERENCES TO CHAPTER 5

(1) GB 1136922   (DOW CORNING CORP)
(2) GB 1141551   (DOW CORNING CORP)
(3) GB 1141840   (DOW CORNING CORP)
(4) FR 1564841   (DOW CORNING CORP)
(5) FR 1568920   (DOW CORNING CORP)
(6) F. GALASSO *et al.*, Appl. Phys. Letters q (1) 37-39 (1968)
(7) R.P.I. ADLER *et al.*, Appl. Phys. Letters 13 (1) 16-19 (1968)
(8) US 3416951   (R.L. HOUGH)
(9) FR 1583684   (ECOLE NATIONALE SUPERIEURE DES MINES DE PARIS ET
                    S.N.E.C.M.A.)
(10) FR 1505474   (CIE FRANCAISE THOMSON HOUSTON)
(11) J.B. HIGGING *et al.*, J. ELECTROCH. SOC. 116 (1) 137-43 (1969)
(12) US 3398013   (J.J. KROCHMAL)
(13) US 3294880   (SPACE AGE MATERIALS)
(14) FR 1511672   (SPACE AGE MATERIALS)
(15) GB 998089   (T.I. GROUP SERVICES LTD)
(16) US 3269802   (HORIZONS INC)
(17) US 3433725   (R.L. HOUGH)
(18) US 3370923   (R.L. HOUGH)
(19) US 3429722   (THE CARBORUNDOM CO)
(20) FR 1551091   (UNITED AIRCRAFT CORP)
(21) GB 1177782   (UNION CARBIDE CORP)
(22) GB 1159210   (UNION CARBIDE CORP)
(23) US 3428719   (FMC CORP)
(24) US 3565749   (FMC CORP)
(25) FR 2002846   (CELANESE CORP)
(26) GB 1204622   (GENERAL TECHNOLOGIES CORP)
(27) DE 1949128   (NATIONAL RESEARCH DEVELOPMENT CORP)
(28) DE 1944504   (CIE FRANCAISE THOMSON HOUSTON)
(29) FR 2170952   (ARMINES)
(30) US 3811927   (GREAT LAKES CARBON CORP)
(31) US 3868230   (AVCO CORP)
(32) FR 1598323   (ECOLE NATIONALE SUPERIEURE DES MINES DE PARIS ET
                    S.N.E.C.M.A.)
(33) FR 2337214   (AVCO CORP)
(34) US 3549413   (GENERAL TECHNOLOGIES CORP)
(35) US 3607367   (GENERAL TECHNOLOGIES CORP)
(36) US 3668006   (GENERAL ELECTRIC CO)
(37) V.G. SAMOILENKO, Chem. Abstr., 87 (14) 215-105/786 k (1977)
(38) GB 2014972   (THE SECRETARY OF STATE FOR DEFENCE, LONDON)
(39) FR 2036618   (CIE FRANCAISE THOMSON-HOUSTON)
(40) US 4107352   (WESTINGHOUSE CANADA LTD)
(41) GB 2080781   (AVCO CORP)
(42) J. ECONOMY *et al.*, Appl. Polym. Symp., n° 29, 105-115 (1976)
(43) R.Y. LIN *et al.*, Appl. Polym. Symp. n° 29, 175-188 (1976)

(44) US 3668059   (THE CARBORUNDUM CO)
(45) US 3620780   (THE CARBORUNDUM CO)
(46) US 3634132   (LOCKHEED AIRCRAFT CORP)
(47) GB 1307133   (THE CARBORUNDUM CO)
(48) US 3630766   (THE CARBORUNDUM CO)
(49) J. ECONOMY *et al.*, Appl. Polym. Symp. n° 21, 130-140 (1973)
(50) US 3728168   (NATIONAL RESEARCH CORP)
(51) US 3640693   (UNITED AIRCRAFT CORP)
(52) JP 7348720   (AGENCY OF INDUSTRIAL SCIENCES AND TECHNOLOGY)
(53) J. SUMNER, ENGINEER, n° 7, p 69 (1974)
(54) EP 32097   (SOC. EUROPEENNE DE PROPULSION)
(55) GM GAVRILOV *et al.*, Chem. Abstr. 93 (26) 281-243.945m (1980)
(56) US 4162301   (UNION CARBIDE CORP)
(57) FR 2064410   (BAYER AG)
(58) FR 2133771   (BAYER AG)
(59) DE 2500082   (UKAEA)
(60) GB 1535471   (TOYO BOSEKI K.K.)
(61) US 4104445   (MONSANTO CO)
(62) US 4175153   (MONSANTO CO)
(63) US 4222977   (MONSANTO CO)
(64) EP 47640   (MONSANTO CO)
(65) US 3850689   (UNITED AIRCRAFT CORP)
(66) R.W. CAHN, NATURE 260 (37) 11-12 (1976)
(67) S. YAJIMA *et al.*, J. Am. Cer. Soc. 59 (7-8) 324-27 (1976)
(68) FR 2308590   (THE RESEARCH INSTITUTE FOR IRON, STEEL AND OTHER
                    METALS OF THE TOHOKU UNIVERSITY)
(69) FR 2308650   (THE RESEARCH INSTITUTE FOR IRON, STEEL AND OTHER
                    METALS OF THE TOHOKU UNIVERSITY)
(70) FR 2345477   (THE RESEARCH INSTITUTE FOR IRON, STEEL AND OTHER
                    METALS OF THE TOHOKU UNIVERSITY)
(71) DE 2236078   (BAYER AG)
(72) FR 2197829   (BAYER AG)
(73) FR 2190764   (BAYER AG)
(74) JP 81110733 (UBE UNIVERSITY LTD)
(75) S. YAJIMA, Chem. Abstr. 94 (20) 285-161.317q (1981)
(76) EP 51855   (UBE INDUSTRIES LTD)
(77) GB 2021545   (DOW CORNING CORP)
(78) US 4298559   (DOW CORNING COPR)
(79) US 4298558   (DOW CORNING CORP)
(80) US 4310651   (DOW CORNING CORP)
(81) GB 2081286   (DOW CORNING CORP)
(82) GB 2081288   (DOW CORNING CORP)
(83) GB 2081289   (DOW CORNING CORP)
(84) JP 78103025 (TOKUSHU MUTI ZAIRY KENKYUSHO)
(85) W. WORTHY, C & EN, june 9, 1980, p. 20
(86) JP 7881727   (TOKUSHU MUKI ZAIRYO KENKYUSHO)
(87) JP 7984000   (ASAHI CHEMICAL INDUSTRY CO)
(88) EP 21844   (UBE INDUSTRIES LTD)
(89) EP 23096   (UBE INDUSTRIES LTD)
(90) EP 30105   (UBE INDUSTRIES LTD)
(91) EP 48957   (UBE INDUSTRIES LTD)
(92) FR 1598321   (ECOLE NATIONALE SUPERIEURE DES MINES DE PARIS)
(93) US 3971840   (THE CARBORUNDOM CO)
(94) US 4075276   (THE CARBORUNDUM CO)
(95) WO 81/02733 (KENNECOTT CORP)
(96) WO 81/02734 (KENNECOTT CORP)
(97) WO 81/02755 (KENNECOTT CORP)

(98)   US 4097294   (R.W. RICE)
(99)   EP 55076   (UBE INDUSTRIES LTD)
(100)   US 4340619   (DOW CORNING CORP)

CHAPTER 6

Other Fibres

1. Review

Besides the inorganic fibres and whiskers mentioned above, other fibres of considerable interest were glass fibres and related products e.g. rock and slag wool and natural fibres e.g. asbestos. These fibres however are outside the scope of this monograph and therefore will not be reviewed herein. Until recently the development of interesting inorganic fibres other than those quoted above were not taken into consideration by fibre manufacturers.

2. New Types of Fibres developed since 1970

In the late seventies a few fibres serving special purposes have been developed. VERSAR INC. (1) used the "relic" process (see Chapter 4, page 37) for the preparation of calcium fluoride fibres : organic rayon precursor fibres are impregnated with a calcium nitrate solution and subsequently immersed into a solution of ammonium fluoride to thereby precipitate calcium fluoride within the precursor fibres; after washing and drying the fibres are heated to drive off the organic material and to sinter the calcium fluoride fibres. These fluoride fibres are suitable to reinforce hydrofluorocarbon elastomers. In another development calcium sulfate or gypsum whisker fibres have been prepared, particularly as a substitute for asbestos fibres; fibrous calcium sulfate has been produced for example by autoclaving a dilute aqueous suspension of gypsum to provide for discrete whisker crystal development (2-4) and is advantageously employed as reinforcement fibre; in order to prevent rehydratation, the fibres have been stabilized by applying organic (5, 6) or mineral (7) coatings.

*　　*

*

REFERENCES TO CHAPTER 6

(1)   US 4104395   (VERSAR INC)
(2)   US 4152408   (CERTAIN-TEED CORP)
(3)   JP 77156198   (KANEGAFUCHI KAGAKU KOGYO K.K.)
(4)   JP 77156199   (KANEGAFUCHI KAGAKU KOGYO K.K.)
(5)   DE 2702097   (BAYER AG)
(6)   DE 2702100   (BAYER AG)
(7)   DE 2738415   (SUDDEUTSCHE KALKSTICKSTOFFWERKE AG)

CHAPTER 7

Monocrystalline Fibres;
Growth Techniques

1. Summary of the Prior Art Growth Techniques

Monocrystalline fibres, also called whiskers, were reported as early as 1574.
They became the subject of intensive study in the 1950s with the discovery
of the high strength of single crystal fibres. The ultrahigh strength of
monocrystalline fibres is attributed to their crystalline perfection, without
the defects responsable for the lower strength of polycrystalline fibres.
HARDY (1) presented in 1956 a comprehensive review of studies of monocrystalline fibres during the past two hundred years. Additional reviews of single
crystal growth followed, including those by COLEMAN (2), and LEVITT (3).
Monocrystalline fibres have been grown by a variety of techniques, which
include growth from the vapor phase, growth from solutions, growth from melts,
and electrolytic growth. A fascinating technique for low-cost production of
monocrystalline fibres was discovered in 1965 by WAGNER (4) : the vapor-
liquid-solid growth (VLS growth).
TYCO LAB., INC. (Waltham. Mass.) introduced in 1967 the growth of continuous
single crystal filaments of aluminium oxide from the melt using the so-called
edge-defined film-fed growth technique (EFG growth).
In the 1960s, UNITED AIRCRAFT RESEARCH LAB. developed a new attractive
solidification technique of eutectic alloys, solving in one single growing
step, spatial distribution and bonding problems in making monocrystalline
fibre-reinforced composites.

2. Developments since 1970

The fascinating properties of monocrystalline fibres have stimulated
extensive research since 1970. The ultrahigh strength of monocrystalline
fibres, which is about double those attainable in polycrystalline fibres, has
still a rapidly growing potential in fibre strengthening metal, ceramic and
plastic matrices.

More recently great activity has been demonstrated in the optical, magnetic
and electronic applications of monocrystalline fibres.
Highly significant effort has also gone forward in the technique of
unidirectional solidification of eutectic or eutectic-type melts which
technique will be treated in Chapter 3 of Part II.
Remarkable progress has been made in the continuous filament technology.
This chapter is intended to provide an understanding of growth techniques in
relation with the fibre material. Indications of applications are also
presented.

2.1. Growth from the vapor phase.   Fibre growth from the vapor phase has become an important method for monocrystalline fibre production. This growth method can be subdivided in two major growth techniques : the evaporation-condensation technique, and the chemical reaction technique. More than thirty elements and hundred compounds have been grown in fibre form with these techniques.

2.1.1. Evaporation-condensation.   This technique involves sublimation or evaporation of a fibre material source, mass transport through the vapor phase, and condensation at the growth site under low supersaturations (commonly established by a controlled temperature gradient).
FURUTA (5)(6) prepared tellurium whiskers by the sublimation of metallic tellurium on the surface of platinum grid at various temperatures.
In another experiment he observed root growth of the whiskers when using a platinum grid on which thallium had been evaporated as an impurity (7).
GAIDUKOV (8) grew antimony filamentary crystals on the walls of an ampoule which was in the 390-440°C temperature range.
Whisker growth of aluminium has been studied by BLECH (9) by heating at 350°C a thin aluminium film, covered with a layer of TiN.
Filamentary bismuth crystals have been grown by GAIDUKOV (10) by condensing evaporated bismuth on a truncated stainless steel cone.
In a method of making a light polarizing material, SLOCUM (11) grew a plurality of metal (Ag, Au, Cu, Al) whiskers parallel to a smooth transparent surface in the direction of the vapor stream.
GLASS (12) prepared graphite and Al_2O_3 elongated fibres by condensing from the vapor state while electrostatically charging the fibres to fan out the ends of the fibres so that the ends are separated and individually exposed to vapor. By depositing simultaneously metal from a gaseous metal mixture on the fibres as they are being formed, metal clad fibres can be obtained which are useful for making reinforced metal articles (13).

CdSe and ZnO needles, grown under an electric field influencing the growth direction along the c axis, are reported by YOSHILE (14).

ZnS needles were grown from the vapour phase in a sealed vessel with a temperature gradient, by SHICHIRI (15).

IWANAGA (16) reports the growth of CdS needle crystals in an atmosphere containing excess Cd or S vapor, by the sublimation method. The dominant growth direction in excess Cd vapor was <0001>, and in excess S vapor <000$\bar{1}$>. He also studied the effects of an electric field on the CdS needle growth (17).

The growth of needles of $WO_{2.9}$ by heating small grains of WO_3 using the illumination of an intense electron beam in an electron microscope, is described by BONNET (18).

2.1.2. Chemical Reaction.   This technique comprehends a complex field of various chemical reaction systems, including reduction of compounds, thermal decomposition of gaseous compounds, discharge induced reaction of gaseous species, chemical transport processes, etc.

OKUYAMA developed an electrical discharge-induced decomposition process of gaseous metal compounds for the growth of metallic (W, Mo, Cr) whiskers (19-23). This process, also called "cathodic whisker growth" involves the decomposition of metal carbonyl vapors by applying a high electric potential, the whiskers being formed on the cathode of a vacuum diode.
Similar results have been obtained by LINDEN (24).

The growth of crystalline needles of W by reaction of tungsten oxide with acetone has also been reported by OKUYAMA (25).

The growth of Cu whiskers by hydrogen reduction of copper iodide is described by GOTOH (26).

A new form of Ge whiskers has been grown on a gold alloyed germanium substrate surface by thermal decomposition of GeH_4 by MIYAMOTO (27).

FAERMAN (28) prepared Ge whiskers by pyrolysis of tetrabutylgermanium in making activated field-effect emitters.

The growth of acicular crystals of Ge by chemical transport of Ge with H_2O is reported by SCHMIDT (29). Ge whiskers have also been grown by TATSUMI (30) on gold plated germanium substrates by thermal decomposition of GeH_4.

Controlled growth of α-Fe whiskers of various orientations by hydrogen reduction of ferrous halides is reported by GARDNER (31). BOJARSKI (32) and KANEKO (33) studied the growth of Fe whiskers by reduction of ferrous chloride with hydrogen.

AHMAD (34) prepared uniaxial Fe whiskers of uniform high quality by passing hydrogen over a halide of iron while applying a magnetic field perpendicular to the reaction zone.

HOLLANDER (35) grew graphite whiskers on a rhodium covered substrate by thermal decomposition of a hydrogen vapor.

NICKL (36) reported a method of producing filamentary monocrystals of Si by contacting vaporous $SiCl_4$ with a mist of Si droplets. This method is analogously applicable for producing whiskers of metals and oxides.

The growth of needles of Si and other semiconductor materials (GaAs, InSb, etc.) by vapor transport with a subsulfide forming agent (H_2S), is described by HUML (37).

Hollow whiskers of Si grown by pyrolysis of alkyl silane compounds on poly-crystalline quartz substrates are reported by AVIGAL (38).

The plasmochemical synthesis of needle-like diamond, using aliphatic hydrocarbons with hydrogen as the reactive gaseous mixture, is reported by MANIA (39).

TOMITA (40) prepared whiskers of metals (W, Mo) or carbides (SiC, B_4C) by reaction of a dispersion of reactable materials in the presence of a catalyst.

Cu-Ni alloy whiskers were grown by hydrogen reduction of a CuI-NiBr2 mixture, by HAMAMURA (41). Some aspects of the growth of TiN, ZrN, TiC and ZrC whiskers by reacting metal chlorides with hydrogen or CH_4, are studied by KATO (42).

Whiskers and hollow crystals of Cr_5Si_3 were grown on a silicon plate from a gas mixture of chromium dichloride, hydrogen and argon in a temperature range of 950-1150°C, by MOTOJIMA (43).

Needle crystals of Cd Ge P_2 have been prepared by chemical transport with iodine and PCl_3, by MIOTKOWSKI (44).

MOTOJIMA (45) reported the chemical vapor growth of LaB_6 whiskers from a gas mixture of $LaCl_3$, BCl_3, H_2 and Ar on a graphite substrate at 1100-1300°C.

CrB hollowed-pillar crystals were also grown from a gaseous mixture of $CrCl_2$, BCl_3, H_2 and Ar in a temperature range of 1050-1100°C, by MOTOJIMA (46).

The growth of needle-like crystals of SnO_2 by reaction of Sn with water vapor, has been described by HATANO (47).

A method for needle-like crystal growth of In_2O_3, by reaction of an equimolar mixture of In_2O_3 and SnO_2 powders with 6-15% graphite powder in a nitrogen atmosphere, is reported by SHIMADA (48).

Needle crystals of anatase (TiO_2) grown by chemical transport reactions with HBr and HCl, are studied by IZUMI (49).

Whiskers of $MgAl_2O_4$ have been prepared by heating magnesia and alumina at a temperature of 1300-1800°C in the presence of a carbonaceous material (50).

Zn_2GeO_4 needle crystals grown at 1000-1170°C by simultaneous oxidation of Zn and GeO vapor species (which were continuously generated by the reduction of Zn_2GeO_4 powder with graphite), are reported by ITO (51).

GRIMSHAW (52) produced Al_2O_3 whiskers by contacting a mixture of alumina and carbon with gaseous aluminium trihalide.

A continuous process for the production of Al_2O_3 whiskers by reaction of aluminium vapour with hydrogen and water vapor is described by KELSEY (53).

High strength Al_2O_3 whiskers manufactured by passing water vapor containing hydrogen and aluminium chloride over aluminium or its alloys, are reported by MINAGAWA (54).

GENERAL ELECTRIC developed some processes for producing Al_2O_3 whiskers, based on the reaction of heated aluminium and silica in an atmosphere of hydrogen or an inert gas (55, 56) or by passing hydrogen and water vapor over a charge of molten aluminium (57, 58, 59).

LONZA described methods for producing SiC whiskers from gaseous reactant species (60) in the presence of carbon or on graphite coated substrates (61, 62).

Single crystal needles of SiC have also been grown by KANEGAFUCHI BOSEKI KABUSHIKI KAISHA, by reacting a silicon containing material, a halogenated substance and carbon (63, 64).

LEE (65) studied the preparation of SiC whiskers by reacting, at an elevated temperature silica, elementary carbon, a source of sulphur, hydrogen and a gaseous source of carbon.

Silicon nitride whisker growth by reaction of a silicon/silica mixture with nitrogen, is reported by EVANS (66).

Whisker growth of carbides, nitrides and borides of Ti or Zr by subjecting a Ti or Zr halide to an electrical discharge in the presence of a gas of the group consisting of nitrogen, ammonia, a hydrocarbon and a boron halide, is described by TAKAHASHI (67).

HUML (68) grew AlN whiskers by contacting a subvalent aluminium compound (AlBr, $AlCl_2$, etc.) in a gaseous state with a nitriding agent.

The production of B_4C whiskers by reacting vaporous boric oxide (B_2O_3) with a hydrocarbon gas, is reported by CLIFTON (69).

KATO (70) prepared whiskers from the $TiCl_4$ - CH_4 - H_2 system.

TiC_x (x = 0,47 - 1,0) whiskers were produced by the reaction of methane, $TiCl_4$ and hydrogen on a carbon-alumina coated tungsten wire heated to 1200-1350°C, by HAMAMURA (71).

The growth of TiN needles from the $TiCl_4$ - H_2 - N_2 system, has been studied by KATO (72).

Needle-shaped crystals of MoO_2, obtained at high temperatures (1000-800°C) by transport with iodine, are reported by BERTRAND (73).

The growth of NiO whiskers by chemical transport of NiO powder with HCl gas, is described by SAITO (74).

The growth of ZnO needle crystals by the oxidation of zinc vapor produced by the reduction of ZnO powder with carbon powder, has been studied by MATSUSHITA (75). ZnO needle-shaped crystals grown by the chemical reaction of ZnF_2 with water vapor, is reported by SHIBATA (76).
Growth of ZnO whiskers obtained by reaction of Zn vapor in the presence of oxygen gas, has been observed by SHARMA (77).

The growth and electrical properties of needles of CdO provided by reacting Cd vapor with oxygen, has been examined by HAYASHI (78).

YOSHIDA (79) grew needle crystals of Zn_2SnO_4 using a starting mixture of ZnO powder and Sn powder.

The manufacture of whiskers of BeO and Cr_2O_3 by oxidation of a metal source or by the reduction of an oxide in a reducing atmosphere of hydrogen or hydrocarbon, in which potassium chloride and traces of iodine or sodium chloride and traces of iodine are added to the reaction mixture, is described by THOMSON HOUSTON (80).

GaAs whisker growth by thermal decomposition using AsH_3 and trimethylgallium, is reported by KASAHARA (81).

MOTOJIMA (82, 83) grew ZrP whiskers by reaction of $ZrCl_4$ and PCl_3, and TiP whiskers by reaction of $TiCl_4$ and PCl_3.

2.2. Growth from Solutions.   The growth of needle-like crystals of $CaSO_4$. $2H_2O$ by inoculating stable supersaturated calcium sulfate solutions with dry seed crystals, has been investigated by NANCOLLAS (84).

Single crystal needles of $Ca_{10}(PO_4)_6$ $(OH)_2$ (hydroxyapatite) have been grown by the hydrothermal method using $Ca(OH)_2$ as flux and employing the oscillating temperature technique, as described by EYSEL (85). Temperature and pressure were oscillated from 750 to 880°C and from 14500 to 17500 psi, respectively.

CHENOT (86) reported the conversion of brushite ($CaHPO_4.2H_2O$) to needle-like monetite ($CaHPO_4$) by control of the pH and temperature of starting brushite slurries.

BRIDENBAUGH (87) developed a process for growing single crystal fibres of rare earth pentaphosphates for waveguiding laser-type applications by mixing oxides of rare earths with a H_3PO_4 solution and controlling the rate of metaphosphoric acid conversion by introducing water vapor into the reaction zone.

Neodymium pentaphosphate (NdP_5O_{14}) needles for laser applications were grown from phosphoric acid solutions containing alkaline metals, Sb and Pb admixtures, by LITVIN (88).

EBERL reported the manufacture of calcium sulfate hemihydrate whiskers by heating an aqueous solution of calcium sulfate dihydrate to a temperature of 110-150°C (89). The resulting whiskers may be treated to form a protective coating on the whiskers (90).

BAYER AG (91, 92) has been investigating the growth of calcium sulfate anhydrites or hydrates.

EDINGER (93) investigated the factors which affect size and growth rates of the habit faces of gypsum, the degree of supersaturation playing a major role.

TiO_2 acicular crystals produced from a calcined blend of TiO_2, salt and oxyphosphorus components, are described by BERISFORD (94).

A method for preparing whiskers of β-PbO_2 by heating an intermediate lead oxide in an aqueous alkali hydroxide solution under a pressure exerted by use of oxygen or an oxygen-containing gas, has been described by TORIKAI (95).

$K_2Ti_6O_{13}$ (potassiumhexatitanate) fibres were synthesized from the system $TiO_2.n H_2O$ - KOH - H_2O at 390°C, as reported by T. OOTA (96).

Needle-like crystal growth of α-MnO_2 from starting materials of γ-MnOOH and KOH, has been studied by YAMAMOTO (97).

The hydrothermal growth of tobermorite ($Ca_6Si_6O_{18}.4 H_2O$) needle crystals under various temperatures and CaO/SiO_2 ratios was examined by HAMID (98).

Willemite (Zn_2SiO_4) needle-like crystals were grown from HCl solutions under hydrothermal conditions, as reported by KODAIRA (99).

The growth of needle-like crystals of silver amalgam in aqueous solution on a mercury surface by means of one kind of internal electrolysis, is described by NANEV (100).

YAMAMOTE (101) reported the growth of needles of $CuGaS_2$ and $CuGa_{1-x}In_xS_2$ by using In as a solvent for a Cu, Ga, Sn containing solute.

2.3. Growth from gels.   This method is particularly useful for materials which cannot readily be grown by other methods. The gel medium (usually silica gel) prevents convection currents or turbulence and, by remaining chemically inert, provides a good medium in which the crystal nuclei are delicately held in position, allowing smooth growth.

The whisker growth of SnI_2 by the gel method has been reported by DESAI (102). A controlled reaction between $SnCl_4$ and KI by a diffusion process in a silica gel medium has been used. At higher concentration of KI solutions, only SnI_4 is obtained (103)(104).

SHIOJIRI (105) grew TlI elongated crystals by diffusion of NaI and $TlNO_3$ solutions through silica hydrogels.

A new gel method for growing large needles and single crystals of $PbCl_2$ using a two-stage chemical reaction in the gel, has been reported by ABDULKHADAR (106). In the first stage, one of the nutrients is incorporated in the gel as a colloidal precipitate (lead tartrate) and the other nutrient (HCl) is allowed to diffuse into the gel to produce the crystals in the second stage. He obtained $PbBr_2$ needles by using KBr as the other nutrient (107). In a further experiment he prepared needles of ZnS and PbS by direct interaction between colloidal suspensions in gels (108). $(NH_4)_2S$ and H_2S have been used as the source of sulphur, and $ZnCl_2$ or Pb $(CH_3COO)_2$ have been used as the source of the metal.

$(Sr_xBa_{1-x})CO_3$ (x = o to 1) crystals in the form of multiple-thread screws, called "braids", are observed by GARCIA-RUIZ (109) during experiments with various gel pH and different $BaCl_2$ or $SrCl_2$ solutions.

Filamental crystals of calciumiodate $(CaIO_3)$ grown by injecting sodium iodate solution into a silica gel and feeding calcium chloride solution above it, are reported by JOSHI (110).

BONCHEVA-MLADENOVA described the growth of elongated crystals of Ag_2SeO_4 in silica gels, the chemicals used were $AgNO_3$ and H_2SeO_4.

Whiskers of Ag_3PO_4 have been grown in silica gel by interdiffusion of aqueous Na_2HPO_4 and $AgNO_3$ solutions and by buffering of an acid Ag_3PO_4 solution in gel containing sodium acetate, as reported by MENNICKE (111).

COPY (112) made a comparison study of $CaSO_4.2H_2O$ grown in clay gels and in sodium silicate gels. He obtained elongated crystals in sodium silicate gels at 40°C and pH 7,5.

Using organic gels, BANKS (113) prepared crystals of several alkaline earth orthophosphates. He observed needle-like growth of $MgNH_4PO_4.6$ H_2O and $CaNH_4PO_4.8$ H_2O.

2.4. Growth from the Melt.   Because the VLS (vapor-liquid-solid) growth technique involves a peculiar growth mechanism, this technique will be treated separately.

2.4.1. From melt solutions using solvents or fluxes.   Whiskers of NiO, MgO and solid solutions thereof produced by reacting a heated mixture of nickel and/or magnesium salt (chloride) and a suitable auxiliary salt with water vapor, are described by BRUBAKER (114, 115).

FUJIKI (116) prepared crystalline fibrous potassium titanate $(K_2O.6TiO_2)$ by melting a mixture of a titanium component (titanium oxide) and a potassium component (potassium carbonate), quenching the molten product and removing the excess K_2O.

Needles of β-Si_3N_4 grown by reaction of a Si melt with a Si_3N_4 crucible, and removing the unreacted Si are reported by INOMATA (117).

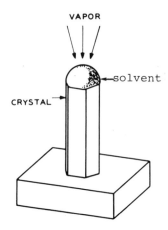

Idealized drawing of VLS mechanism illustrated for growth of a whisker

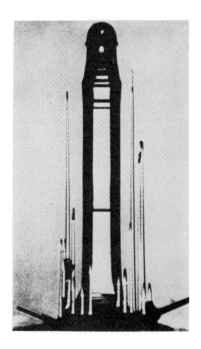

VLS growth of whiskers

TiO_2 whiskers prepared by reaction of titanium compounds at about 600°C with a melt of one or more oxyacid salts (Li_2SO_4, Na_2SO_4, etc.) are reported by FARBENFABRIKEN BAYER (118). Analogous reactions can be used for the production of oxydes of other metals (Be, Mg, Zn, Al, etc.) (119).

Undoped and Ga-doped $ZnSiP_2$ needlde-like crystals grown by spontaneous crystallization from a melt containing nine parts zinc or tin and one part $ZnSiP_2$ under a slow cooling rate, are described by SIEGEL (120).

Rodlike crystals of mixed (La, Eu, Y, Ce, Ba, Cs) hexaborides for thermoionic emission, grown from an aluminium flux using oxide powders, are reported by OLSEN (121).

DESAI (122) observed the growth of whiskers of fluorite (CaF_2) using alkali halides as the flux.

Whisker growth of $PbWO_4$ employing Na_2WO_4 as the flux has been observed by ARORA (123).

Whiskers of In_5S_4 using molten tin as the solvent, have been obtained by WADSTEN (124).

The growth of needles of $CuGaS_2$ and $CuGa_{1-x}In_xS_2$ in In solution has been studied by YAMAMOTO (125).

The Vapor-Liquid-Solid (VLS) Growth Technique.   This technique, in which vapor, liquid, and solid phases are involved is characterized by a liquid growing tip site on the growing fibre. The liquid becomes, during the growth, supersaturated with material supplied from the vapor, while precipitation occurs at the solid-liquid interface.

Extensive studies and experiments have been carried out for the VLS Si whisker growth, which differ mainly in the choice of the solution or liquid-forming agent.

WEYHER (126, 127) used Pt and Au as the liquid-forming agent, while VAN ENCKEVORT (128) studied the use of Sn and Al experimenting with a tin or aluminium coated graphite substrate, on which the metal upon heating contractted to droplets. Au has also been used as solvent by WAGNER (129, 130). Fe has been reported as the liquid-forming agent by CONRAD (131).

GIVARGIZOV (132) studied the use of Au, Pt, Ag, Ni, Ga and Au-Ga.

DIEPERS (133) developed an efficient process for whisker growth of Si and III-V compounds, proposing as the liquid-forming agent Au, Pt, Pd, Ni, Cu or Ag for the Si growth, and a compound partner (Ga, In, etc.) for the III-V compound growth. Ga droplets are also reported by ARTHUR (134) for the growth of GaAs.

GIVARGIZOV (135) experimented with gold as the liquid-forming agent for the growth of III-V compounds.

Pillar crystal growth of BP in the presence of fine molten Ni particles on a substrate, has been reported by MOTOJIMA (136).

The growth mechanism of Cu whiskers by the hydrogen reduction of cuprous

iodide, has been studied by NITTONO (137) which observed cuprous iodide
droplets at the whisker tips.

AHMAD (138) prepared B whiskers by the reduction of BBr_3 and BCl_3 with hydrogen using Au, Pt, Cu and Ag as liquid forming impurities.

PHILIPS (139, 140, 141) developed some methods for the VLS growth of SiC
whiskers proposing Fe, Au, Fe-Al alloy, etc. as the liquid-forming agent.

A method for growing MgO whiskers with Al, Ca, Si as the liquid-forming
impurity, has been reported by BURNETT (142).

Lead has been used as a suitable solvent for the VLS growth of $\alpha-Al_2O_3$
whiskers, as reported by GRIBKOV (143).

DZIUBA (144) grew SbSI whiskers, postulating VLS mechanisms of crystallization when using I, S, Sb, Sb_2S_3 and SbI_3 for the starting materials.

The VLS growth of $Sm_4 (SiO_4)_3$ whiskers has been observed by JABLONSKI (145)
when heating a samarium sample in quarz tube filled up with argon at a
pressure of 0,5 atm and at room temperature.

FURATA (146) reported the VLS whisker growth of Te and Thallium-Telluride
using Tl as a liquid-forming agent.

2.4.2. Directly from Melts.   A highly significant effort has been made by
LA BELLE (147, 148, 149) of TYCO LAB. in the production of continuous Al_2O_3
(sapphire) filaments with super strength by using the so called "Edge-Defined
Film-Fed Growth" (EFG) technique. One of the more interesting developments
is the simultaneous production of a plurality of continuous sapphire filaments (150, 151). He proposed also the production of filaments of BeO, Cr_2O_3
and TiO_2 by the same EFG method (152).

ARTHUR D. LITTLE INC. (153) reported a method and apparatus for the growth
of monocrystalline filaments of Al_2O_3, ZrO_2, ThO_2, B and Si, by drawing the
filaments from a molten zone on top of a feed rod. They introduced also an
apparatus for focusing a beam of laser energy around the molten zone (154).

BURRUS (155) described a technique for growing neodymium-doped yttrium
aluminium garnet ($Y_3Al_5O_{12}$) optical fibres by pulling from a melt at one end
of a feeding rod, the melt being heated by a focused laser beam.

The most recent developments can be found in the production of infrared light
transmitting optical fibres.

SUMITOMO ELECTRIC INDUSTRIES (156) developed a process for producing an
infrared light transmitting optical fibre of silver halide, thallium halide,
or a mixed crystal thereof, by drawing a feeding rod into a fibre through a
die. A step-index type fibre is also produced by intimate cladding such
fibres with a crystal layer having a lower refractive index.
They introduced (157) also a process for producing a step-index monocrystalline fibre by drawing a core fibre from a melt through a small tube and cladding with a crystal layer from a second melt. Alkali metal halides are used
for the core and cladding material.

MIMURA (158, 159) proposed a modified floating zone technique to draw monocrystalline fibres for infrared optical waveguides. This technique is
applicable to NaCl, KCl, KBr, LiF, CaF_2 and BaF_2.

OKUMARA (160) reported a CsI optical fibre grown from the melt using the modified pulling down method.

Optical fibres comprising a monocrystalline core and cladding by crystal pulling from melts, are described by NAKAGOME (161). The core material may be a halide of an alkali metal, or a halide or oxide of an alkaline earth metal.

BRIDGES (162) reported single-crystal optical fibres of AgBr grown at the exit of a nozzle fed by liquid AgBr.

Single-crystal fibre waveguides of thallium halides (TlCl, TlBr) silver halides (AgCl,AgBr) and alkali halides (KBr, KCl, CsI), grown in capillaries wetted by the molten starting material from where the fibres were drawn using a roller, are reported by VASIL'EV (163).

2.5. Other Techniques.   An infrared fibre of AgCl clad AgBr core fabricated by an extrusion process, is described by ANDERSON (164).

Iron monocrystals of acicular form grown by electrolytic deposition from iron-containing acid aqueous solution, are reported by UVAROV (165).

Rectangular hollow needles of NbO_2 obtained by electrodeposition from molten fluorides containing significant amounts of oxide contamination, are prepared by COHEN (166).

BRANDIS (167) reported a technique for growing whiskers which utilizes electromigration of metal. For example, when adjacent sections of Al and an aluminium alloy and Cu are subjected to an electric current, whiskers grow from the aluminium region.

A peculiar technique for the production of Al_3Ni filaments, which involves selectively removing the Al matrix phase from a undirectional solidified eutectic Al-Al_3Ni composite alloy to expose the second filament phase, has been described by QUINLAN (168) and HUSSEY (169). They propose for the removal of the matrix phase the use of a leaching oxalic acid - hydrogen chloride solution, respectively an aluminium halide containing molten salt electrolyte.

A similar technique has been reported by ELLIS (170) and LEMKEY (171) for the production of surfaces composed of protruding aligned fibres, by removing the matrix phase to a desired depth.

*   *

*

REFERENCES TO CHAPTER 7

(1)   H.K. HARDY, Progr. Metal Phys., 6, 1956
(2)   R.V. COLEMAN, "Growth and Properties of Whiskers", Met. Rev. 9,
       35, 1964
(3)   A.P. LEVITT, "Wisker Technology", Wiley-Interscience (JOHN WILEY AND
       SONS, New York, 1970
(4)   R. WAGNER and W.C. ELLIS, Trans. Met. Soc. AIME, 233, 1057, 1965
(5)   N. FURUTA et al., Japan J. Appl. Phys., 11, 8, 1113 - 1118, 1972
(6)   N. FURUTA et al., Japan J. Appl. Phys., 14, 7,   929 - 934, 1975
(7)   N. FURUTA et al., Japan J. Appl. Phys., 13, 3,   545 - 546, 1974
(8)   Y.P. GAIDUKOV et al. Instrum. and Exp. Techn., 19, 2, 569 - 570, 1976
(9)   I.A. BLECH et al. J. Crystal Growth, 32, 2, 161 - 169, 1975
(10)   Y.P. GAIDUKOV et al. Instrum. and Exp. Techn., 15, 5, 1565 - 1566, 1972
(11)   US 3969545 and 4049338, R.E. SLOCUM
(12)   US 3615258, J.P. GLASS
(13)   US 3915663 and 3536519, J.P. GLASS
(14)   T. YOSHILE et al., J. Crystal Growth, 51, 3, 624 - 626, 1981
(15)   T. SHICHIRI et al., J. Crystal Growth, 43, 3, 320 - 328, 1978
(16)   H. IWANAGA et al., J. Crystal Growth, 50, 2, 552 - 554, 1980
(17)   H. IWANAGA et al., J. Crystal Growth, 49, 3, 541 - 546, 1980
(18)   J.P. BONNET, S. HORIVCHI et al., J. Crystal Growth, 56, 3, 633 - 638,
       1982
(19)   F. OKUYAMA et al., Appl. Phys. Lett., 35, 1, 6 - 7, July 1974
(20)   F. OKUYAMA et al., Appl. Phys., 22, 1, 83 - 87, 1980
(21)   F. OKUYAMA et al., Appl. Phys., 22, 1, 39 - 46, 1980
(22)   F. OKUYAMA, Appl. Phys. Lett., 36, 1, 46 - 47, 1980
(23)   F. OKUYAMA, J. Crystal Growth, 49, 3, 531 - 540, 1980
(24)   DE 2615523, H. LINDEN
(25)   F. OKUYAMA, J. Appl. Phys., 46, 8, 3255 - 3259, 1975
(26)   Y. GOTOH, Japan J. Appl. Phys., 11, 10, 1403 - 1412, 1972
(27)   Y. MIYAMOTO et al., Japan J. Appl. Phys., 14, 9, 1419 - 1420, 1975
(28)   V.I. FAERMAN et al., Instruments and Experimental Techniques, 17, 5,
       1405 - 1406, 1974
(29)   G. SCHMIDT et al., J. Crystal Growth, 55, 3, 599 - 610, 1981
(30)   Y. TATSUMI et al., J. Phys. Soc. Japan, 47, 5, 1511 - 1517, 1979
(31)   R.N. GARDNER, J. Crystal Growth, 43, 4, 425 - 432, 1978
(32)   Z. BOJARSKI et al., J. Crystal Growth, 46, 1, 43 - 50, 1979
(33)   T. KANEKO, J. Crystal Growth, 44, 1, 14 - 22, 1978
(34)   US 3607451, I. AHMAD
(35)   US 3664813, E.F. HOLLANDER
(36)   US 3607067, J. NICKL
(37)   US 3519492, J.O. HUML et al.
(38)   Y. AVIGAL et al., J. Crystal Growth, 26, 1, 157 - 161, 1974
(39)   R. MANIA et al., Crystal Research and Technology, 16, 7, 785 - 788,
       1981
(40)   US 3840647, C. TOMITA et al.

(41) K. HAMAMURA *et al.*, J. Crystal Growth, 46, 6, 804 - 806, 1979
(42) A. KATO, J. Crystal Growth, 49, 1, 199 - 203, 1980
(43) S. MOTOJIMA *et al.*, J. Crystal Growth, 55, 3, 611 - 613, 1981
(44) I. MIOTKOWSKI *et al.*, J. Crystal Growth, 48, 4, 479 - 482, 1980
(45) S. MOTOJIMA *et al.*, J. Crystal Growth, 44, 1, 106 - 109, 1978
(46) S. MOTOJIMA *et al.*, J. Crystal Growth, 51, 3, 568 - 572, 1981
(47) T. HATANO *et al.*, J. Crystal Growth, 57, 1, 197 - 198, 1982
(48) S. SHIMADA *et al.*, J. Crystal Growth, 55, 3, 453 - 456, 1981
(49) F. IZUMI *et al.*, J. Crystal Growth, 47, 139 - 144, 1974
(50) GB 1511393, S.A. SUVOROV *et al.*
(51) S. ITO *et al.*, J. Crystal Growth, 47, 310 - 312, 1979
(52) US 3947562, R.W. GRINSHAW *et al.*
(53) US 3658469, R.H. KELLSEY
(54) US 3582271, S. MINAGAWA *et al.*
(55) US 3505014, GENERAL ELECTRIC
(56) GB 1190283, GENERAL ELECTRIC
(57) GB 1203342, GENERAL ELECTRIC
(58) GB 1203343, GENERAL ELECTRIC
(59) US 3668062, GENERAL ELECTRIC
(60) FR 2097792, LONZA
(61) FR 2150396, LONZA
(62) FR 2091412, LONZA
(63) FR 2087892, KANEGAFUCI BOSEKI KABUSHIKI KAISHA
(64) FR 2075819, KANEGAFUCI BOSEKI KABUSHIKI KAISHA
(65) US 3709981, S.A. LEE *et al.*
(66) US 3677713, C.C. EVANS
(67) US 3657089, T. TAKAHASHI *et al.*
(68) US 3598526, J.O. HUML *et al.*
(69) US 3525589, R.A. CLIFTON
(70) A. KATO *et al.*, J. Crystal Growth, 37, 3, 293 - 300, 1977
(71) K. HAMAMURA *et al.*, J. Crystal Growth, 26, 2, 255 - 260, 1974
(72) A. KATO *et al.*, J. Crystal Growth, 29, 1, 55 - 60, 1975
(73) O. BERTRAND *et al.*, J. Crystal Growth, 35, 325 - 328, 1976
(74) S. SAITO *et al.*, J. Crystal Growth, 30, 1, 113 - 116, 1975
(75) T. MATSUSHITA *et al.*, J. Crystal Growth, 26, 1, 147 - 148, 1974
(76) N. SHIBATA *et al.*, Japan, J. Appl. Phys., 11, 6, 775 - 779, 1972
(77) S.D. SHARMA *et al.*, J. Appl. Phys., 42, 13, 5302 - 5304, 1971
(78) S. HAYASHI, Rev. Elect. Com. Labor. 20, 7 - 8, 698 - 709, 1972
(79) R. YOSHIDA *et al.*, J. Crystal Growth, 36, 1, 181 - 184, 1976
(80) GB 1190038, THOMSON HOUSTON
(81) J. KASAHARA *et al.*, J. Crystal Growth, 38, 1, 23 - 28, 1977
(82) S. MOTOJIMA *et al.*, J. Crystal Growth, 30, 1, 1 - 8, 1975
(83) S. MOTOJIMA *et al.*, J. Electrochem. Soc., 123, 2, 290 - 295, 1976
(84) G.H. NANCOLLAS *et al.*, J. Crystal Growth, 20, 2, 125 - 134, 1973
(85) W. EYSEL *et al.*, J. Crystal Growth, 20, 3, 245 - 250, 1973
(86) US 3927180, C.F. CHENOT
(87) US 4002725, P.M. BRIDENBAUGH *et al.*
(88) B.N. LITVIN *et al.*, J. Crystal Growth, 57, 3, 519 - 523, 1982
(89) FR 2179760, J.J. EBERL *et al.*
(90) US 3961105, J.J. EBERL *et al.*
(91) DE 2752367, BAYER AG
(92) FR 2377970, BAYER AG
(93) S.E. EDINGER, J. Crystal Growth, 18, 3, 217 - 224, 1973
(94) US 3728443, R. BERISFORD *et al.*
(95) US 3959453, E. TORIKAI *et al.*
(96) T. OOTA *et al.*, J. Crystal Growth, 46, 3, 331 - 338, 1979
(97) N. YAMAMOTO *et al.*, Japan J. Appl. Phys., 13, 4, 723 - 724, 1974

(98)   S.A. HAMID, J. Crystal Growth, 46, 3, 421 - 426, 1979
(99)   K. KODAIRA et al., J. Crystal Growth, 29, 1, 123 - 124, 1975
(100) C. NANEV et al., Kristall und Technik, 10, 4, 355 - 360, 1975
(101) N. YAMAMOTO et al., Japan, J. Appl. Phys., 11, 9, 1383 - 1384, 1972
(102) C.C. DESAI et al., J. Crystal Growth, 51, 3, 457 - 460, 1981
(103) C.C. DESAI et al., J. Crystal Growth, 50, 2, 562 - 566, 1980
(104) C.C. DESAI et al., J. Crystal Growth, 53, 2, 432 - 436, 1981
(105) M. SHIOJIRI et al., J. Crystal Growth, 43, 61 - 70, 1978
(106) M. ABDULKHADAR et al., J. Crystal Growth, 48, 1, 149 - 154, 1980
(107) M. ABDULKHADAR et al., Crystal Research and Technology, 17, 1,
       33 - 38, 1982
(108) M. ABDULKHADAR et al., J. Crystal Growth, 55, 2, 398 - 401, 1981
(109) J.M. GARCIA-RUIZ et al., J. Crystal Growth, 55, 2, 379 - 383, 1981
(110) M.S. JOSHI et al., Kristall und Technik, 15, 10, 1131 - 1135, 1980
(111) S. MENNICKE et al., J. Crystal Growth, 26, 2, 197 - 199, 1974
(112) R.D. CODY et al., J. Crystal Growth, 23, 4, 275 - 281, 1974
(113) E. BANKS et al., J. Crystal Growth, 18, 2, 185 - 190, 1973
(114) US 3875296, B.D. BRUBAKER
(115) US 3711599, B.D. BRUBAKER
(116) US 4265872, Y. FUJIKI
(117) Y. INOMATA et al., J. Crystal Growth, 21, 2, 317 - 318, 1974
(118) FR 2080633, FARBENFABRIKEN BAYER
(119) FR 2078869, FARBENFABRIKEN BAYER
(120) W. SIEBEL et al., Kristall und Technik, 15, 8, 947 - 954, 1980
(121) G.H. OLSEN et al., J. Crystal Growth, 44, 3, 287 - 290, 1978
(122) C.C. DESAI et al., J. Crystal Growth, 44, 5, 625 - 628, 1978
(123) S.K. ARORA et al., J. Crystal Growth, 57, 2, 452 - 455, 1982
(124) T. WADSTEN, J. Crystal Growth, 52, 673 - 678, 1981
(125) N. YAMAMOTO et al., Jap. J.of Appl. Phys. 11, 9, 1383 - 1384, 1972
(126) J. WEYHER, J. Crystal Growth, 43, 2, 235 - 244, 1978
(127) J. WEYHER, J. Crystal Growth, 43, 2, 245 - 249, 1978
(128) W.J.P. VAN ENCKEVORT et al., J. Electrochem. Soc., 128, 1,
       154 - 161, 1981
(129) US 3493431, R.S. WAGNER et al.
(130) US 3505127, R.S. WAGNER et al.
(131) US 3607054, R.W. CONRAD
(132) E.I. GIVARGIZOV, J. Crystal Growth, 20, 3, 217 - 226, 1973
(133) US 4155781, H. DIEPERS
(134) US 3635753, J.R. ARTHUR et al.
(135) E.I. GIVARGIZOV, Kristall und Technik, 10, 5, 473 - 484, 1975
(136) S. MOTOJIMA, J. Crystal Growth, 49, 1, 1 - 6, 1980
(137) O. NITTONO et al., J. Crystal Growth, 42, 175 - 182, 1977
(138) I. AHMAD et al., J. Electrochem. Sy, 118, 10, 1670 - 1675, 1971
(139) GB 1400562, PHILIPS
(140) GB 1213867, PHILIPS
(141) GB 1213156, PHILIPS
(142) US 3630691, P. BURNETT et al.
(143) V.N. GRIBKOV et al., Soviet powder metallurgy and metal ceramics,
       11, 4, 264 - 267, 1972
(144) Z. DZIUBA, J. Crystal Growth, 35, 3, 340 - 342, 1976
(145) L. JABLONSKI et al., J. Crystal Growth, 57, 1, 206 - 108, 1982
(146) N. FURATA et al., Jap. J. Appl. Phys., 11, 11, 1753 - 1754, 1972
(147) H.E. LABELLE et al., Mat. Res. Bull., 6, 7, 571 - 580, 1971
(148) H.E. LABELLE et al., Mat. Res. Bull., 6, 7, 581 - 590, 1971
(149) US 3627574, H.E. LABELLE
(150) US 3953174, H.E. LABELLE
(151) H.E. LABELLE, J. Crystal Growth, 50, 1, 8 - 17, 1980

(152) US 3650703, H.E. LABELLE
(153) US 3944640, ARTHUR D. LITTLE INC.
(154) US 4058699, ARTHUR D. LITTLE INC.
(155) US 4040890, C.A. BURRUS
(156) EP 0056996, SUMITOMO ELECTRIC IND.
(157) EP 0056262, SUMITOMO ELECTRIC IND.
(158) Y. MIMURA et al., Jap. J. Appl. Phys., 19, 5, L 269 - L 272, 1980
(159) FR 2424552, Y. MIMURA et al.
(160) Y. OKUMARA et al., Jap. J. Appl. Phys., 19, 10, L 649 - L 651, 1980
(161) GB 2016731, Y. NAKAGOME et al.
(162) T.J. BRIDGES et al., Optics Letters, 5, 3, 85 - 86, 1980
(163) A.V. VASILE'V et al., Sov. J. Quantum Electr., 11, 6, 834 - 835, 1981
(164) US 4253731, R.H. ANDERSON
(165) US 3496078, L.A. UVAROV et al.
(166) U. COHEN, J. Crystal Growth, 46, 1, 147 - 150, 1979
(167) E.K. BRANDIS et al., IBM Techn. Discl. Bull., 12, 10, 1544, 1970
(168) US 4191561, K.P. QUINLAN et al.
(169) US 4100044, C.L. HUSSEY et al.
(170) DE 1964991, C.H. ELLIS et al.
(171) US 4209008, F.D. LEMKEY et al.

PART II

INORGANIC FIBRES COMPOSITE MATERIALS

<u>NOTES ON CONTENTS</u>

1° *This part deals only with entirely inorganic materials.*

2° *Convential composite materials such as asbestos or glass fibre reinforced concrete, are not discussed here.*

3° *Irrespective of their composition, all composite materials obtained by the unidirectional solidification of eutectic melts and unidirectional decomposition of eutectoid materials are discussed in Chapter 3, page 132.*

CHAPTER 1

Metal Matrix Composites

1. Summary of the Prior Art

Most of the inorganic fibrous materials known today have been developed
during the sixties and although they were not always available at that time
on an industrial scale, research on their use in the production of high
grade composite materials started immediately.
In particular, lightweight, high-temperature resistant materials and
corrosion-resistant refractory materials were badly needed by the aeronautic
and nuclear industries for the construction respectively of gas-turbines,
rocket engines, etc. and reactors of ever increasing power.

At the end of the sixties a great number of possible combinations of fibres
and metal matrices had already been investigated. Table 1 gives a summary of
the most representative combinations which have been the subject of intensive
study. However it was soon realized that only a few of these combinations
had any engineering potential. During fabrication of metal matrix composites
by the classical techniques, the fibres come in contact with the metal at
high temperature, and the reactions occuring between the fibres and the
matrix often destroys the strength of the fibres to such an extent that no
reinforcement of the matrix at all is obtained.
Mainly responsable for the fibre deterioration were phenomena such as :
intermetallic diffusion between fibres and matrices with subsequent formation
of new low strength phases, dissolution of the fibre in the matrix or chemical
reaction between the same. Typical examples of the latter are the formation
of carbides respectively borides at the fibre-matrix interface in the case
of carbon respectively boron reinforced metals such as Ti, Ni, or Cr base
alloys. Also problematic, although to a lower degree, were poor wetting, low
interfacial bonds and unbridgeable differences in thermal expansion
coefficients between fibre and matrix.
Many attempts have been made to circumvent the afore mentionned problems;
new molding techniques such as electroplating, plasma-spraying or evaporating
the matrix material on to the fibres did not require the high temperature
molding phase; precoating the fibres with appropriate layers protected them
against surface reactions or confered better wettability or bonding characteristics. But these improvements also raised dramatically the production
costs of the metal matrix composites.

TABLE 1: Summary of the Prior Art Metal Matrix Composites.

Fibres	Matrix	Fabrication method
Be	Ag, Ag-Al-Ge Al	powder met. extrusion
Be, Ni-B	Be, Ni-B	melting, direc. solid.
Bronze, steel	Pb	-
Cr, Al$_3$Ni	Al-Cu, Cu-Cr	melting
Cu$_6$Sn$_5$	Cu-Sn	direct. solid.
NiAl$_3$	Al-Ni	extrusion, melting
Mo	Ti-6Al-4V Ti	cold press. sinter, extrude
Steel	Al	hot press.
Steel	Al-10% Si	casting
Steel	Ag	vac. infil.
Steel	Ag	-
Steel	Al	diffusion bonding
Steel	Mg	casting
Steel, W	Al alloy	hot rolling
Steel, Mo, W and Ge	Ag, Cu, Al	slip cast, vac. infil., extr., swaging
W	Cu	liq. infil.
W	Cu, Al	liq. infil.
W, Mo	Ni, Co, L605 Nichrome steel	powder met. forge, roll
W	Cu alloys L605, Ni-Fe	vac. infil.
W	Ni	pneumatic impaction
W	U	vac. infil.
W	Ni, Ti, Al, Cu	molecular forming
W	W	vapor deposition
W	Cu-Ni	-
Al$_2$O$_3$	Ag	liq. infil.
Al$_2$O$_3$	Al	liq. infil.
Al$_2$O$_3$	2-S and Al alloys	casting
Al$_2$O$_3$	Ni and Ni alloys	liq. infil. powder met., electroforming

FIBRES	MATRIX	FABRICATION METHOD
Al_2O_3	Fe, Ti, Nichrome	powd. met., melting
Al_2O_3	refr. metals	slurry
B_4C	Fe-Ni, Al	-
Boron	Al, Cu, Ni	casting, powd. met.
Boron	Al alloy	liq. infil.
Boron	Nickel	electroforming
Boron	Al	powder met. plus laminated sheets
Boron	Ni, Fe, Ti Cu	powd. met., diffusion bonding
Boron	Ti	diffusion bonding
E glass, also SiO_2-MgO-Al_2O_3	Cu, Brass Steel	liq. infil.
Fiberfrax	Ni-Sn	vac. infil.
Glass	Al, Al alloys	-
Glass + UO_2	Al alloys	rolling
Graphite	Al-4%-Ni-Cr	powder met.
NbC	Nb	direct solid
SiC	Ni-Cr	direct solid
SiC	Al	-
Si_3N_4	Ag, Ag + 1% Si	hot press
SiO_2	Al	hot press
ZrO_2, Y_2O_3, HfO_2, ThO_2, HfB_2, etc.	W	extrusion

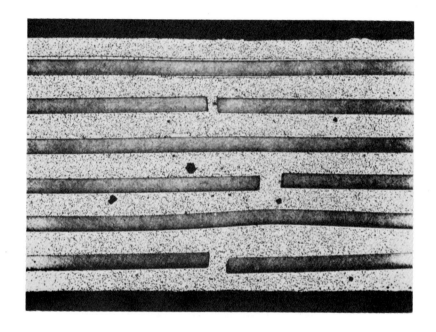

Steelreinforced aluminium composite

2. Developments since 1970

Notwithstanding the first discouraging results with metal matrix composites, many new research projects were started and a great number of technical publications were issued during the last decade. From these publications it appears that the efforts were more selectively directed to the development of a limited number of specific combinations.

For the sake of clarity the results of the recent literature study have been subdivided according to the three main combinations of fibre-metal matrix composites, *i.e.* metal-metal; carbon, boron, carbide, boride-metal; ceramic oxide or glass-metal.

2.1. The metal fibre - metal matrix systems.   Table 2 summarizes the most important combinations that have been further investigated since 1970. In the light metal matrix systems the beryllium fibre reinforced composites seems the most promising, in particular the beryllium-titanium matrix combination.

Another class in the metal-metal systems which has retained great attention, is based on refractory Ni, Co or Cr base superalloys; reinforcement of these metals by tungsten or molybdenum fibres however mostly failed as a result of excessive interdiffusion reactions. The most promising composites in this kind of superalloy-series are now obtained by the unidirectional solidification method (see Chapter 3).

TABLE 2: Composites of the Metal Fibre/Metal Matrix System

FIBRES	MATRIX	FABRICATION METHOD	FIELD OF APPLICATION	AUTHOR	REF.
Al$_3$Co$_2$	Al-1,5 Cu alloy	directional solidification	as electrical conductors of high tensile strength	UNITED TECH-NOLOGIES	(9)
Al$_4$Ce	Al-Ce alloy 8 - 12% Ce	directional solidificaton	nuclear industry	PECHINEY	(10)
Stainless steel	Al	powder metallurgy; lamination	aircraft industry	TUMANOV $e.a.$ PECHINEY	(14) (21)
Be ribbons	Al; Ti	Al or Ti cladded Be rods are inserted in drilled Al or Ti preforms; mechanical deformation of the preform	shafts for high speed rotating engines e.g. rotor shafts in heli-copter, controlrods in in aircraft	R. SCHMIDT	(5)
Ta	Mg	infiltration technique	aircraft industry	I. AHMAD $e.a.$	(22)
Mo	Ti or Ti alloy	powder metallurgy; fibre alignment by extrusion, rolling, etc.	supersonic aircraft rocket propulsion	A. SCHWOPE $e.a.$	(2)
Be or Ti cladded Be	Ti alloy; pure Ti	explosive welding of interposed layers of Ti and Be fibres	aircraft construction	R. TRABOCCO	(4)
Be	Ti Ti-6Al-4V Ti-6Al-6V2Sn Ti-5Al-2,5Sn	hot extrusion of mixed precursor of Ti and Be; the latter can be present as fibre preforms	aircraft industry	BRUSH WELLMANN INC.	(13)
Ni$_3$Al	Ni - 2 to 10% Al	powder metallurgy; the fibre like phase is formed in situ during the mechanical working	oxidation resistant, high strength at high and low temperature	CABOT CORP.	(27)

IF-D

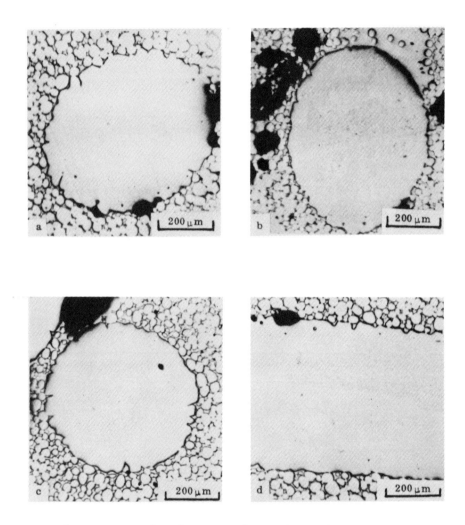

Microstructure of W-Ni-Fe composites reinforced with W fibres

TABLE 2 (Cont.)

FIBRES	MATRIX	FABRICATION METHOD	FIELD OF APPLICATION	AUTHOR	REF.
Ni$_3$Ta	Ni - Ta alloy	unidirectional solidification		EURATOM	(32) (34)
Ni Cr Al Y	Ni alloy	powder metallurgy	sealing elements in turbines, compressors	BRUNSWICK CORP.	(37)
W/1% ThO$_2$	superalloy	investment casting	jet engines	US ARMY	(74) (75)
W	W-Ni-Fe alloy	liquid phase sintering; the w fibres are recrystallized to avoid dissolution		E. ZUKAS NASA	(38) (40)
Stainless steel	Ni alloys	electroforming	rocket engines	NASA	(39)
Mo, Ti, Nb	Ni superalloys	powder metallurgy		US ARMY MATERIEL DEVELOPMENT	(41)
W	Cu	melt impregnation	electrical machinery	HITACHI	(59)
W filament coated with successive layers of B and Ti or Al$_2$O$_3$	Fe or alloys	not defined		US COMPOSITES CORP.	(63)
Ti or Mo band coated with Sic	Ti, Mg	various processes		ARMINES	(61)
Nb filaments	Ni, Cu, Ag	Nb filaments inbedded in a Cu, Ni or Ag matrix are passed through a molten bath of Sn, until formation of Nb$_3$Sn	super-conductors	IMPERIAL METAL IND.	(64)

2.2. The carbon, boron, carbide, boride .... fibre-metal matrix system.
This is by far the most extensively investigated class of metal base
composites; carbon, boron, ... fibres have indeed a unique combination of
properties such as refractoriness, high strength to weigth ratio, and are now
commercially available in continuous filament or whisker form.
Furthermore they can now effectively be protected against interface reaction
with the metal matrix by the use of appropriate coatings.
Table 3 gives an extended summary of the results achieved up to now.

In the light weight composite sector, the carbon fibre reinforced aluminium
or magnesium combinations present the best prospects both by their relatively
simple methods of manufacture and by their low cost and availability.
Boron fibres coated with silicium carbide (BORSIC) are also advantageously
used with Al or Mg matrices.

Silicon carbide fibres and filaments as well as whiskers on the other hand
have predominantly been used as the reinforcing phase in the nickel, cobalt,
and other refractory metal base composites.

Carbon fibres were also inbedded in softer metals such as Cu, Ag, bronzes,
etc. which were used as bearing or friction elements.

Surface Fracture of carbon fibre reinforced Al

TABLE 3 : Composites of the Carbon, Boron, Carbide, ... Fibre/Metal Matrix System

FIBRES	MATRIX	FABRICATION METHOD	FIELD OF APPLICATION	AUTHOR	REF.
BORSIC	Al	powder metallurgy; the composite article is clad-ded with a sheet of Ti by diffusion bounding	turbine blades	UNITED AIRCRAFT PRATT & WHITNEY	(12) (19) (23) (20)
C coated with a Ag-Al alloy	Al-4,5M-0,6Mn-1,5Mg	powder metallurgy	high temperature resistant components; arms, aircraft	UNION CARBIDE	(17)
continuous B or BORSIC + β SiC discon-tinuous fibres	Al	liquid fase hot pressing		R. HERMANN *e.a.* F. SWINDELS	(24) (26)
C (graphite, amorphous carbon)	Ni/Co Aluminide	coating C fibres with Ni or Co; mixing with Ni Co Al powder; hot pressing		UNITED TECH-NOLOGIES	(44)
C coated with boride of Ti, Zr, Hf	Al or Al alloys Mg; Pb; Sn; Cu; Zn	melt impregnation		FIBER MATERIALS INC.	(46)
C pretreated with molten NaK alloy	Al	melt impregnation		A.P. LEVITT ANVAR	(47) (53)
C	Al alloy con-taining carbideforming metal e.g. Ti, Zr, ...	melt impregnation		HITACHI	(56)
SiC with W core	Al-Cu alloy	coating the filaments with Cu; passing the Cu coated filaments through an Al melt		THOMSON CSF	(65)
C	Mg or Mg alloy	hot pressing of alternating layers of matrix metal and fibres; small amounts of Ti, Cr, Ni, Zr, Hf or Si are added to promote wetting	Aircraft industry	A.P. LEVITT	(43)
C	Mg or Mg alloy	melt impregnation, the mol-ten Mg matrix contains small amounts of magnesium nitride to enhance wetting of the fibres	turbine fan blades; pressure vessels; armor plates	I. KALNIN	(45)
C coated with Ti	Mg	melt infiltration; liquid phase hot pressing		A.P. LEVITT	(58)
SiC	Be or alloys with Ca, W, Mo, Fe, Co, Ni, Cr, Si, Cu, Mg, Zr	vacuum impregnation with molten Be or plasma spraying fibres with Be and consoli-dation by metallurgical process	aerospace and nuclear industry	RESEARCH INST. FOR IRON & STEEL OF THE TOHOKU UNIVERSITY	(8)
B + stainless steel; Borsic + Mo fibres	Al Ti	impregnation; spraying, etc. combination of high strength ductile and brittle fibres	aerospace industry	INST. FIZIKI FVERDOGO TELA AKADEMII NAUK SSSR	(16)
SiC	Ti or alloy Ti-3Al-2,5V	hot pressing of interposed layers of fibres and matrix sheets; SiC fibres are previously coated with zirconium diffusion barrier layer	compressor blades; airfoil surfaces	GENERAL MOTORS	(3)

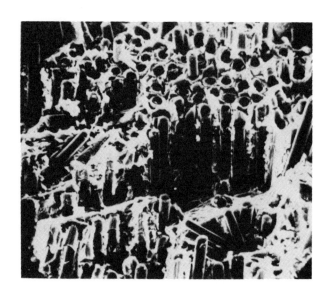

Fracture Surface of Magnesium composite reinforced with Ti cladded carbon fibres

TABLE 3   (End)

FIBRES	MATRIX	FABRICATION METHOD	FIELD OF APPLICATION	AUTHOR	REF.
Carbides of Nb, Ta, W	Ni-Co alloys Fe-Cr alloys	unidirectional solidification	aircraft industry	ONERA GENERAL ELECTRIC	(31) (36) (33) (35)
SiC containing 0,01-20% free carbon	Cr base alloys	powder metallurgy; the free carbon reacts with the chromium to form carbides, thus improving bonding ability	high strength heat resistant material e.g. vanes and blades for turbines; rocket nozzles	THE RESEARCH INSTITUTE FOR SPECIAL INOR-GANIC MATERIALS, JAPAN	(28)
SiC, containing 0,01-30% free carbon	Co or Co base alloys	powder metallurgy or melt impregnation; carbide formation between the fibres and the Co matrix	high strength heat resistant material e.g. vanes and blades for turbines; rocket nozzles	THE RESEARCH INSTITUTE FPR SPECIAL INOR-GANIC MATERIALS, JAPAN	(29)
SiC containing 0,01-20% free carbon	Mo base alloys	powder metallurgy	high strength heat resistant material e.g. vanes and blades for turbines; rocket nozzles	THE RESEARCH INSTITUTE FOR SPECIAL INOR-GANIC MATERIALS, JAPAN	(30)
C coated with carbides	Ni or Ni alloys	melt impregnation	aeronautic industry	UNION CARBIDE	(51) (52)
B	Cu-Ti-Sn alloy	liquid phase sintering	cutting tools	GENERAL DYNAMICS	(62)
C	bronze	various processes	bearing materials	UK ATOMIC ENERGY AUTH.	(55)
C	Cu alloy	powder metallurgy; the fibres are mixed with a slurry of Cu powder and 2% of a carbide-forming metal powder (Ti, Cr, ..)	high strength electric conductive materials	HITACHI	(57)
C coated with Ti boride	Al, Cu, Sn, Pb, Ag, Zn, Mg	the matrix contains alloying elements of Ti and B to prevent deterioration of the TiB coating of the fibres	aeronautic industry	AEROSPACE CORP.	(48)
C coated with Ni	metals with melting point lower than that of Ni	melt impregnation		BROWN-BOVER NASA	(49) (50)
C coated with SiO_2 + SiC	Al, Mg, Ti, Ni	melt impregnation, powder metallurgy		FIBER MATERIALS	(54)
Monocarbides of Ta, Ti, W	Al, Al-Si alloy, Ag or Ag alloys Cu or Cu alloys	melt impregnation	abrasion resistant materials	UNION CARBIDE	(66)
SiC	Si	melt impregnation		GENERAL ELECTRIC S. YAJIMA	(67) (70)
β SiC	Ag or Ag alloys		electric conductors, contacts, ...	THE RESEARCH INSTITUTE FOR IRON, STEEL AND OTHER METALS OF THE TOHOKU UNIVERSITY	(68)
C	Si	powder metallurgy	abrasive materials	GENERAL ELECTRIC	(69)
SiC whiskers	Ag	hot pressing (in zero-gravity conditions)		S. TAKAHASHI	(72)
C coated with TiB	Mg, Pb, Zn, Cu, Al, Zn	melt impregnation		FIBER MATERIALS	(42)

2.3. <u>The ceramic oxide or glass fibre - metal matrix system</u>.   Ceramic oxides
show a pronounced reactivity versus numerous metals at higher temperature,
while most glass fibres present relatively low softening points.
This kind of fibre has therefore only been investigated in combination with
low melting metals.
Combination of Al_2O_3(sapphire) and ceramic $Al_2O_3.SiO_2$ fibres with Al or Mg
matrices are characteristic for this class of composites.
A summary is given in Table 4.

TABLE 4: Composites of the Ceramic Oxide or Glass Fibre - Metal Matrix System

FIBRES	MATRIX	FABRICATION OF METHOD	FIELD OF APPLICATION	AUTHOR	REF
Al_2O_3; SiC; Aloxy-nitride	Al-Cu alloy	mixing minute filaments in molten matrix; after solidification, the filaments penetrate through grain boundary regions		R. ROSENBERG	(1)
Al_2O_3	Al-Li alloy	infiltration with a molten 1-8% Li alloy; reaction occurs between Al_2O_3 fibre and the Li		DU PONT DE NEMOURS	(6)
Al_2O_3-SiO ; Y Al_2O_3	Al Al-Zn alloy	melt impregnation; powder metallurgy, etc.	aeronautics	SUMITOMO	(15) (26) (73)
Ni coated glass ceramic fibres	Al	powder metallurgy	sliding parts	HONDA MOTOR CO	(25)
Al_2O_3 continuous and poly-crystalline	Mg or alloy	melt impregnation of aligned fibres preferred composites contain at least 50% of fibres	turbine blades, spings, shafts, ...	DU PONT DE NEMOURS	(7)
C,B, glass, ceramic, metal fibres	Mg alloy	powder metallurgy; the composite contains two different Mg alloy phases	multiple applications	DANNÖHL	(11)
glass fibres	light metals	melt impregnation	constructional parts	FA W. SCHADE	(18)
metal coated glass fibres	Fe, Be, Ti, Al, Sn	chopped metal coated glass fibres are mixed with glass filaments and metal powder; hot pressing	dimensional stable machine parts, e.g. friction elements	OWENS CORNING	(60)
glass fibres	Pb	melt impregnation	battery plates; bearing materials; acoustic insulation	D.M. GODDARD e.a.	(71)

3. Future Trends and Conclusions

Most of the compatible inorganic fibre-metal matrix combinations have now been extensively investigated with respect to their properties, limitations and potential applications.
In the near future the activity of the R & D centers will probably shift towards improvements of the existing fabrication processes or to the development of new methods for the manufacture of articles with specific configuration of the fibre strengthening phase. In this respect the uni-directional solidification method seems to offer the most promising prospects (see Chapter 3, page 126).

It is curious to ascertain that the Japanese industry which occupied a leading position in the industrial development of different fibre materials, in particular carbon and carbide fibres, has shown only a modest activity in the field of the metal composites. SUMITOMO did some work in the field of Al_2O_3/Al composites and the recently created foundations RESEARCH INSTITUT FOR IRON, STEEL AND OTHER METALS OF THE TOHOKU UNIVERSITY and RESEARCH INSTITUT FOR INORGANIC MATERIALS show increasing interest in the fibre reinforced superalloys. The lack of an important aeronautic or nuclear industry might be responsible for this phenomenon.

*   *

*

REFERENCES TO CHAPTER 1

(1)   US 3492119   (R. ROSENBERG)
(2)   US 3510275   (A. SCHWOPE)
(3)   US 3717443   (GENERAL MOTORS)
(4)   US 3847558   (R. TRABOCCO)
(5)   US 3938964   (R. SCHMIDT)
(6)   US 4012204   (DU PONT)
(7)   US 4036599   (DU PONT)
(8)   US 4141726   (THE RESEARCH INSTITUT FOR IRON & STEEL)
(9)   US 4148671   (UNITED TECHNOLOGIES CORP.)
(10) FR 1588139   (EURATOM)
(11) FR 2084320   (DANNOHL)
(12) FR 2111243   (UNITED AIRCRAFT)
(13) FR 2196393   (BRUSH WELLMANN INC.)
(14) FR 2242476   (TUMANOV $e.a.$)
(15) FR 2260630   (SUMITOMO)
(16) FR 2416270   (INST. FIZIKI)
(17) DE 1912465   (UNION CARBIDE)
(18) DE 2250116   (FA W. SCHADE)
(19) K KREIDER, "Metal progress" vol 97 (May 1970) n° 5, p 104 - 108
(20) S.A. SATAR $e.a.$, "Journal of Aircraft" vol 8 n° 8 (1971) p 648 - 651
(21) E. ANDERSON $e.a.$, "Revue de Métallurgie" vol 69 n° 2 (Feb. 1972)
      p 165 - 179
(22) I. AHMAD $e.a.$, "Metallurgical Trans." vol 4 (March 1973) p 793 - 797
(23) K. KREIDER $e.a.$, "Metallurgical Trans." vol 4 (April 1973)
      p 1155 - 1165
(24) R. HERMANN, "Navy Technical Disclosure Bulletin" vol 3 n° 6
      (June 1978)
(25) F. SWINDELS, "Amer. Ceramic Sa. Bulletin" vol 54 n° 12 (1975)
      p 1075 - 1078
(26) EP 62496   (SUMITOMO)
(27) US 3715791   (CABOT CORP.)
(28) US 4117565   (CHIAKA ASADA)
(29) US 4147538   (SEISHI YAJIMA)
(30) US 4180399   (CHIAKI ASADA)
(31) FR 2071294   (ONERA)
(32) FR 2084451   (EURATOM)
(33) FR 2136394   (GENERAL ELECTRIC)
(34) FR 2190933   (EURATOM)
(35) FR 2231767   (GENERAL ELECTRIC)
(36) FR 2441665   (ONERA)
(37) GB 1512811   (BRUNSWICK CORP.)
(38) E. ZUKAS $e.a.$, "Journal of Less Common Metals", 32 (1973) p 345 - 353
(39) "NASA TECH BRIEF" June 1974 B74 - 10018
(40) "NASA TECH BRIEF" December 1974 B74 - 10248
(41) NTN - 78/0355

(42)   US 3860443   (FIBER MATERIALS INC.)
(43)   US 3888661   (A.P. LEVITT)
(44)   US 3953647   (UNITED TECHNOLOGIES)
(45)   US 4056874   (CELANESE)
(46)   US 4082864   (FIBER MATERIALS)
(47)   US 4157409   (A.P. LEVITT)
(48)   US 4223075   (THE AEROSPACE CORP.)
(49)   FR 2075256   (BROWN - BOVERI)
(50)   FR 2130603   (NASA)
(51)   FR 2192067   (UNION CARBIDE)
(52)   FR 2192193   (UNION CARBIDE)
(53)   FR 2259916   (ANVAR)
(54)   FR 2323527   (FIBER MATERIALS)
(55)   GB 1403862   (U.K. ATOMIC ENERGY AUTH.)
(56)   DE 2164568   (HITACHI)
(57)   DE 2649704   (HITACHI)
(58)   A.P. LEVITT, "Metallurgical Transactions" vol 3, Sept. 1972
        p 2455 - 2459
(59)   JP 74121734   (HITACHI)
(60)   US 3992160   (OWENS CURNING)
(61)   FR 2080076   (ARMINES)
(62)   US 4116689   (GENERAL DYNAMICS)
(63)   FR 2057466   (US COMPOSITE CORP.)
(64)   FR 2155522   (IMPERIAL METAL INDUSTRIES)
(65)   FR 2165012   (THOMSON CSF)
(66)   FR 2191996   (UNION CARBIDE)
(67)   FR 2227244   (GENERAL ELECTRIC)
(68)   FR 2334757   (THE RESEARCH INSTITUT FOR IRON & STEEL OF THE
                       TOHOKU UNIVERSITY)
(69)   FR 2373348   (GENERAL ELECTRIC)
(70)   DE 2647862   (SEISHI YAJIMA)
(71)   D.M. GODDARD *e.a.*, "Composites" vol 8 n° 2 (April 1977) p 103 - 109
(72)   S. TAKAHASHI, "AI AA Journal" vol 16 n° 5 (May 1978) p 452 - 457
(73)   EP 62496   (SUMITOMO)
(74)   NTN - 79/0176
(75)   PB 81 - 970305

CHAPTER 2

Ceramic Matrix Composites

1. Survey of Known Materials

As early as 1927 a ceramic composite consisting of an alumina matrix reinforced with oxidation-resistant metal wires was patented (1) in order to obtain a material with improved mechanical strength and thermal shock resistance. Already then it was recognized that the thermal expansions of fibre reinforcement and matrix should be matched to achieve strength improvement.

Nevertheless no further developments were reported until the sixties when it became clear that in addition two more conditions had to be met to achieve greater strength increases (2):

- the modulus of elasticity of the reinforcing fibre must be higher than that of the ceramic matrix
- there must be no chemical incompatibility between the fibre and the matrix.

As a result of these restricting conditions, the outlook for metal fibre-ceramic matrix systems was not very promissing with relation to improvement of strength properties (3). In fact thermal shock resistance was the most significant improvement obtained. A summary of fibre-ceramic matrix systems developed in the sixties as well as their properties is given by VASILOS *et al*. (3) and H.W. RAUCH *et al*. (2). Other developments are described in refs. (4 - 12) and are related to methods of producing reinforced ceramic bodies (4 - 8), Mo and Ta fibres in Al_2O_3(9, 10), W fibres in TiC, HfC, TaC and ZrC (9), W and Mo fibres in ZrO_2 and ThO_2 (11) and B, Ta, Ti, W and Mo fibres in carbon (12).

It is striking that only a little work had been done on ceramic fibre-ceramic matrix composite materials at that time, although the conditions mentioned above to achieve strength improvement should be met more easily than with metal fibre - ceramic matrix combinations. This was demonstrated by THE GENERAL ELECTRIC CO (13) : especially strong composite bodies were obtained by incorporating fired threads of alumina in an alumina matrix for example. However further developments of such composites were probably limited due to the lack of suitable ceramic fibres, especially continuous fibres, having improved properties and which had still to be developed (see part I).
In fact other developments were related to ceramic materials reinforced with refractory whiskers such as whisker reinforced Si_3N_4 or ZnO (14), molten calcium aluminates (15) and BN (16), or ceramic materials reinforced with other available fibres such as boron (17) and SiC fibres (18).

Microstructure of magnesium oxide ceramic containing molybdenum metal fibres after hot pressing. 50X.

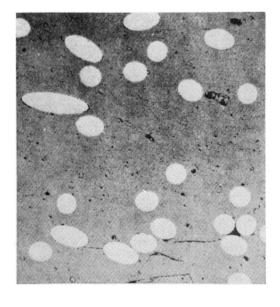

Microstructure of fused silica as hot pressed with molybdenum fibres. 50X.

In the late sixties a new family of ceramic fibres - ceramic matrix composites was introduced which are now generally identified as carbon-carbon
composites. The composites were produced by forming carbon or graphite
fibres into a shape and impregnating the shaped fibres with a carbonizable
resin (19 - 21) or by chemical vapor deposition (CVD) of pyrolytic carbon or
graphite (22); the resin impregnated shape was subsequently carbonized or
graphitized and could be further densified by repeating impregnating and
carbonizing cycles (19) or by CVD of pyrolytic carbon or graphite (20).
The articles thus obtained were strong, light-weight materials and found
many applications in modern industry as will be discussed later.

2. New Materials developed since 1970

2.1. Metal-ceramic systems.   A summary of the metal fibre - ceramic
matrix composites is given in the following table.

TABLE 5: Summary of Metal Fibre - Ceramic Matrix Composites

FIBRES	MATRIX	FABRICATION METHOD	FIELD OF APPLICATION OR SCOPE OF RESEARCH	AUTHOR	REF.
Cr Mo	Al_2O_3-Cr_2O_3 CeO_2 doped Gd_2O_3	hot pressing grains of previously directionally solidified eutectic composites	homogeneous distribution and spacing of thin metal fibres	CLAUSSEN	(23)
Steel, Al alloys	Al_2O_3, fused SiO_2	—	exhaust pipe	TOYOTA MOTOR	(24)
continuous Mo	Al_2O_3	—	bending strength and thermal stability	AKIMENKO	(25)
V,Nb,Ta Cr,Nb,Ta Ta	Cr_2O_3 TiO_2 ZrO_2	directional solidification hot pressing grains of previously grown eutectic	preparation and fracture toughness studies	CLAUSSEN	(26)
Cr	Fe_3O_4,Al_2O_3, Cr_2O_3 and mixtures	directional solidification	gas turbine airfoil	UNITED TECHNOLOGIES	(27)
W, Mo	HfO_2+ZrO_2 and Y_2O_3	—	heat resistance	SHEVCHENKO	(28)
Ta	unstab. HfO_2	directional solidification	solidification behaviour	CLAUSSEN	(29)
W,Mo	stab. HfO_2	hot pressing	throat areas of rocket nozzles	FLETCHER	(30)
W	MgO	—	impact strength	KARPINOS	(31)
Ni,Fe,Co	MgO	hot pressing	strength and fracture toughness	HING	(32)
W	fused SiO_2	hot pressing	mechanical strength	DUNGAN	(36)
Ta,Mo,Nb	UO_2	directional solidification	preparation studies	JEN	(33)
stainless steel	wustite	hot-pressing	fracture stress and toughness	ZWISSLER	(34)
Cu,Cu-Be,Be	Be_4B,Be_2B	hot-pressing, plasma-spraying or vapor deposition	armor material	US ARMY	(35)
Ti,Cr	SiC	whisker formation in situ	preparation study	LUKIN	(37)
Ta,W	Si_3N_4	hot-pressing	gas turbine engines	BRENNAN	(38 - 43)
W,Mo	Si_3N_4	flame spraying silicon and heating in nitriding atmosphere	strength, fracture toughness and fragmentation resistance improvement	NATIONAL RESEARCH DEVELOPMENT	(44)
Mo,Ta,W	Sialon, Si_3N_4, Si_3N_4-C,TaC	hot-pressing	structure reinforced in three mutually perpendicular X, Y and Z directions	AVCO	(45)(46)

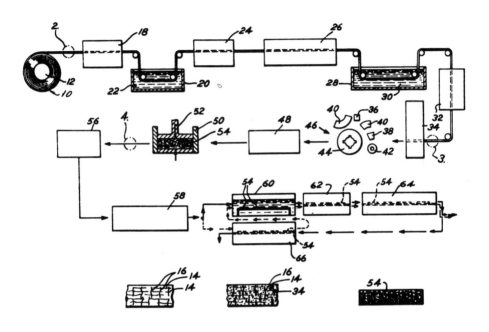

Flow chart of carbon-carbon composite production (THE BENDIX CORP)

Ta	TaC	hot pressing	thermal stress resistance	USAEC	(47)
W,W-Re	TaC	hot-pressing	thermal shock and erosion resistance throat areas of rocket nozzles	FLETCHER	(48)
Nb	MoSi$_2$	hot-pressing	heat conductor	FITZER	(49)
Nb	borosilicate glass	hot-pressing	fine micro-meter-sized fibres; mechanical properties	LUCAS	(207)
Ni	glass-ceramic	hot-pressing	effect of expansion mismatch on mechanical properties	DONALD	(208)
W,Mo, stainless or carbon steel	glass, glass-ceramic	fusing glass coated fibres together using pressure	impressing strength, impact resistance and modulus of elasticity	OWENS-CORNING	(181)
stainless steel	PbO glass	hot-pressing, vacuum injection or pulltrusion	nozzles, nose cones, heat shields, gaskets, seals, armor, electrical conductors or insulators, filters for radiation	OWENS-CORNING	(180)

As can be seen from this summary, developments of metal-fibre-ceramic systems are related to only a limited number of different metal fibres in combination with various ceramic matrices. On account of the difficulties met in producing strong bodies as clarified before, spectacular developments could not be expected. Work on oxide matrices is now focused on directional solidification studies of eutectic compositions (see Chapter 3, page 126).
It is also mentioned here that ceramic materials e.g. fused silica has been reinforced by honeycomb structures of refractory metals (50) and that metal-ceramic composites are also used to produce implantable prostheses (51)(52).

2.2. Ceramic - ceramic systems

2.2.1. Carbon - carbon composites.   As mentioned before, carbon-carbon composites are generally produced by impregnating shaped carbon or graphite fibres with a carbonizable precursor which is subsequently pyrolized or by CVD of carbon or graphite. The resulting materials are strong and lightweight and have excellent high temperature properties; their resistance to corrosion, thermal shock and good electrical and wear characteristics together with their ease to be formed into a wide variety of shapes make them useful for widespread applications (53). As a result carbon-carbon composites are the most developed of the ceramic-ceramic composites.

A summary of these developments is given in the table here under.

TABLE 6 : Summary of Carbon-Carbon Composites

SCOPE OF RESEARCH	FIELD OF APPLICATION	AUTHOR	REF.
torsional strength	internal bone plates	FITZER	(54)
influence of process parameter on mechanical properties	—	FITZER	(55)
co-carbonization of fibres with phenol resin matrices	—	MARKOVIC	(56)

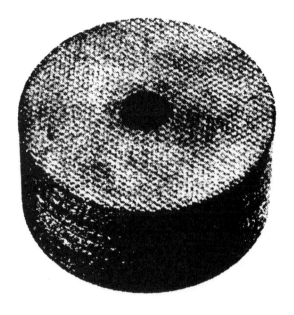

Typical carbon-carbon composite moulding

TABLE 6 (Cont.)

carbon fibre surface treatment; influence on mechanical properties	internal bone plates	FITZER	(57)
fibre-matrix inter-actions	—	BRADSHAW	(58)
optimization of ther-mal shock resistance	heat shield	GUESS	(59)
multidirectional rein-forcement structures	nozzles	SOC. EUR. DE PROPULSION	(60 - 66)
—	friction material for clutches and brakes	EATON CORP.	(67)
bonding wrapped cloth to core	prostheses	FORDATH	(68)
making rigid substrate	several	GOODRICH	(69)
high porosity products	thermal insulation	CONRADTY	(70)
carbonizable mixture contg. carbon bars or needles	electrodes	COAL IND.	(71)
apparatus for continuous CVD	electrodes in fuel cells	PFIZER	(72)
improved oxidation resistance	brake discs	GOODYEAR	(73)
matrix from carbonized coal solution or coal extract	nozzles, cones and bearings	COAL IND.	(74)
mandrel materials; layered products	—	TOHO BESLON	(75)
laminated reinforcement structure	nozzles	SOC. EUR. DE PROPULSION	(76)
laminated reinforcement structure	friction member e.g. aircraft brakes	KULAKOV	(77)
laminated reinforcement structure	brake linings	CEA	(78)
spinning of fibres and matrix; apparatus	nuclear reactors, prostheses	CEA	(79)
selective compression of reinforcement structure	—	SOC. EUR. DE PROPULSION	(80)
ribbon wrapped core	brake discs	GOODYEAR	(81)
non-woven fabric substrates; needle punching	brake discs	DUNLOP	(82)

TABLE 6 (Cont.)

impregnation process	brake discs	CARBORUNDUM	(83)
co-firing novolac fibres and phenol resin matrix	chemistry, nuclear industry, medicine	KANEBO	(84)(85)
laminated reinforcement structure	nozzles, cones, aircraft brakes	CARBONE-LORRAINE	(86)
layered composite, each layer presenting different physical properties	battery electrodes seals, thermal insulation	CONRADTY	(87)
neutron irradiation to improve tensile strength	——	EURATOM	(88)
laminated reinforcement structure	brake discs	MORGANITE MODMOR	(89)
co-firing laminated cotton cloths and resin matrix	chemical, electrical, glass industry and medicine	PHILIPS	(90)
tool for making rigid substrates	——	SOC. EUR. DE PROPULSION	(91)
bonding composite layer to graphite	brake discs	GOODYEAR	(92)(97)
transformation of thermoplastic fibres to a substrate before carbonization	——	DUCOMMUN	(93)
impregnation process	——	SOC. EUR. DE PROPULSION	(94)
continuous filament winding on a mandrel	screws and gaskets	DUNLOP	(95)
bonding composite layer to graphite	friction member	DUNLOP	(96)
needling reinforcement substrate	——	HYFIL	(98)
sheet fabrication	electrodes in fuel cells	KUREHA	(99)
fabrication process of basic substrates	several	DUCOMMUN	(100)
incorporation of a minor quantity of carbon fibres	electrodes	BRITISH NUCLEAR FUELS	(101)
substrate formation	chemistry, foundry, aerospace	SIGRI	(102)
laminated reinforcement structure	brake discs	GOODYEAR	(103)

TABLE 6 (Cont.)

laminated reinforcement structure	brake discs	BENDIX	(104)
circumferentially wound fibres axially compressed	brake discs	CARBORUNDUM	(105-107)
substrate formation	——	UKAEA	(108)
adding fibres to mesophase and graphitizing	electrical and chemical industry	GREAT LAKES CARBON	(109)
CVD impregnation process	——	CEA	(110)
——	electrolytic anode plate	KUREHA	(111)
expanded graphite composite	gaskets	CARBONE-LORRAINE	(112)
SiC coated fibres	——	CARBORUNDUM	(113)
filament winding	pistons for engines	AUDI	(114)
impregnation with mesophase forming binder	——	SCHUNK & EBE	(115)
incorporation of a minor quantity of carbonizable organic fibre	electrodes	NIPPON CARBON	(116)
formation and apparatus	——	NIPPON CARBON	(117)
formation using chopped carbon fibres	several	BITZER	(118)
using pitch fibres	electrodes	KUREHA	(119)(120)
fabrication process; improved interlaminar strength	——	FITZER	(121)
spraying a slurry contg. fibres and binder onto a porous mandrel	aerospace	USAEC	(122)(123)
reinforcement of constructional graphite parts	nuclear reactors	BROWN BOVERI	(124)
——	foundry molts	BICKERDIKE	(125)
radial needling	brake discs	CARBONE-LORRAINE	(126)

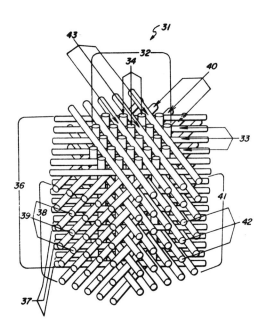

Assembled rigid rods and
infiltration of a geometric
structure thus formed for
the fabrication of a carbon-
carbon composite.
(SCIENCE APPLICATIONS)

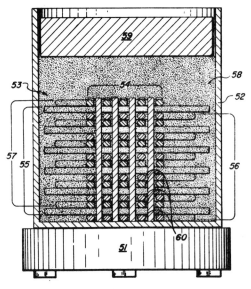

TABLE 6 (Cont.)

extrusion, fibre aligned ribbon; circumferential orientation	brake discs	UNION CARBIDE	(127)
needle-punching	brake discs	DUNLOP	(128)
using organic fibres and petroleum residue precursor materials	electrodes, foundry molds, seals, brake discs	SHELL	(129)(130)
laminated reinforcement structure	brake discs	DUNLOP	(131)
improved chemical erosion resistance; impregnation with molten silicon	aircraft brakes	SECRETARY OF STATE OF DEFENCE; LONDON	(132)
fabrication process; improved interlaminar strength	rocket exhaust systems	UKAEA	(133)(134)
incorporation of a minor quantity of carbon fibres	electrodes	COAL IND.	(135)
spraying superimposed layers onto a former	—	UKAEA	(136)
laminated reinforcement structure	—	UKAEA	(137)
using partially converted PAN fibres	—	NATIONAL RESEARCH DEVELOPMENT	(138)
fabrication process; porous structures	moulds, thermal insulation	FORDATH	(139)(140)
using partly fired fibres; vitreous carbon matrix	—	PLESSEY	(141)(142)
fibres disposed in the same direction	heat exchanger tubes	COURTAULDS	(143)
multidirectional reinforcement structures and impregnation process	aerospace, gas turbines, bioprosthetic devices	SCIENCE APPL.	(144)
three directional reinforcement structures	re-entry vehicles	US SECR. OF AIR FORCE	(145)
woven reinforcement structures; implanting metal and metal carbide wires	ablative materials	FIBER MATERIALS	(146)(147)

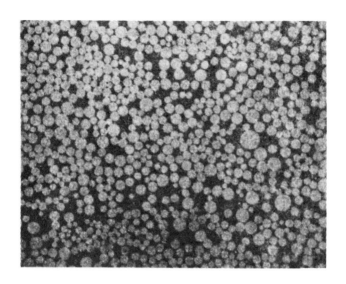

Cross sections of uniaxial and multiaxial silicon carbide fibre reinforced ceramic matrix composite (UNITED TECHNOLOGIES CORP).

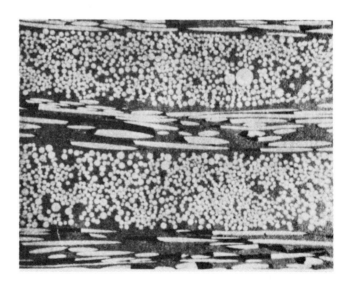

continuous densification process	brakes, missile components	HITCO	(148)(149)
fabrication process; improved interlaminar tensile strength in boron contg. composites	—	HITCO	(150)(151)
fabrication process; vacuum molding; layered products	thermal insulation	US DEP. OF ENERGY	(152)(153)
metallized graphite filament wound structures	aerospace	US DEP. OF ENERGY	(154)
orthogonally woven reinforcement structure	aerospace	GENERAL ELECTRIC	(155)
porous sheet materials	fuel cell electrode substrates	UNITED TECHNOLOGIES	(156)
fabrication process from fibre contg. resin solutions	brakes, seals, aerospace	GREAT LAKES CARBON	(157)
CVD onto filaments; filament wound structures; microcomposite matrix	rocket nozzles, turbine blades, braking elements	ATLANTIC RESEARCH	(158-163)
multidirectional reinforcement structures	ablative materials	FIBER MATERIALS	(164)
heat sink core	friction members for brakes and clutches	GOODRICH	(165)
using chemically extracted and macerated wood precursor fibres	aerospace	HITCO	(166)
carbonizing process	hot pressing molds	FIBER MATERIALS	(167)
multiple heat treatments	—	AVCO	(168)
formation of porous substrates	re-entry systems	HAVEG	(169-171)
needling fibre shapes	—	CARBORUNDUM	(172)
using partially carbonized organic fibres to match volume shrinkage to that of the resin matrix	ablative materials	HITCO	(173)

TABLE 6 (End)

SCOPE OF RESEARCH	FIELD OF APPLICATION	AUTHOR	REF.
isotropic thermal properties; fabrication process	aerospace	USAEC	(174)
conoïd structures	aerospace	USAEC	(175)
overwrapping graphite die wall with a graphite fibre	hot press dies	US SECR. OF THE ARMY	(176)

As follows from the foregoing table it is mainly European (particularly French), and US organisations that are involved in the development of carbon-carbon composites owing in the first place to the development of supersonic aircraft for which light weigth brake systems are needed and the need for high temperature resistant materials in the aerospace and nuclear industries.

2.2.2. Ceramic - glass composites.   The first succesful developments of ceramic fibre reinforced glass and glass-ceramic matrix composites for structural applications were published in the early seventies when several workers (177 - 185) found that some high strength fibres could effectively reinforce lower modulus glass matrices. Boron fibre (178 - 181), carbon fibre (177 - 179)(183)(185), graphite fibre (178)(180)(184) and silicon carbide fibre (178) reinforced glass and glass-ceramics were thus introduced. At the same time, several methods of making these composites were proposed (177 - 182) but the method according to which reinforcing fibres are coated with matrix particles and subsequently consolidated preferably by hot pressing (185) into composite products, has now generally been adopted.

Stimulated by the results obtained, further studies were undertaken mainly dealing with carbon (186 - 188)(209) and graphite (189 - 194) reinforcing fibres. It was recognized that both chemical and mechanical compatibility between carbon fibre and matrix is improved by coating the fibres with SiC or TiC (186 - 188). In another development a new composite from continuous alumina fibres and a glass matrix was introduced (195 - 197); however, graphite fibre reinforced glass systems proved better over all levels of strength and toughness.

As a result of these research programmes UNITED TECHNOLOGIES has recently developed silicon carbide fibre reinforced glass composites (198 - 201) which are claimed to provide a unique combination of both high levels of mechanical performance along with excellent oxidation resistance; the improved properties were obtained by incorporating new types of high temperature fibres based on silicon carbide which have become available on the commercial market and are produced by AVCO and YAJIMA (see part I). Similarly an evaluation of mechanical strength and erosion stability for silicon carbide fibre reinforced quartz glass composites was given by KARPINOS (202).

A study of the properties and structure of the above described composites is given by THOMPSON (203). Applications of specific compositions are given in refs. (193)(194)(197)(201)(203)(204); uses are related to aerospace, internal combustion engines, as well as to domestic cookware and armor applications.

A totally different development is described by CORNING GLASS (205)(206) who patented the production of composite articles comprising a glass or glass-ceramic matrix containing long fibrous single crystals of rutile (TiO_2) of very high aspect ratios grown in situ; the articles are formed by heat treatment of a molten batch of appropriate composition.

2.2.3. Other ceramic - ceramic composites.   Since 1970 workers have shown growing interest both in ceramic whisker and ceramic fibre reinforcement of polycrystalline oxide and non-oxide ceramic matrices. For the sake of clarity these developments have been summarized separately in tables 7 and 8.

TABLE 7 : Summary of Ceramic Whisker - Ceramic Matrix Composites

WHISKERS	MATRIX	FABRICATION METHOD	FIELD OF APPLICATION OR SCOPE OF RESEARCH	AUTHOR	REF
$3Al_2O_3.2SiO_2$, α-Al_2O_3, ZrO_2	oxides and nitrides	hot-pressing	machine components, protective shields	KARPINOS	(210)
$3Al_2O_3.2SiO_2$, α-Al_2O_3, SiC, Si_3N_4, ZnO	TiO_2	hot-pressing	thermal shock resistance	KARPINOS	(211)
$3Al_2O_3.SiO_2$	Al_2O_3, Al_2O_3-Mo, Cr_2O_3, ZrO_2, Al_2O_3-Cr, AlN, BN, Si_3N_4, V_2O_3, TiN, SiO_2	hot-pressing	mechanical and thermal properties	- KARPINOS - TOTSKAYA - GUMENYUK - SAMSONOV	(212 - 223), (231)
α-Al_2O_3, AlN, SiC	$3Al_2O_3.2SiO_2$-Al_2O_3	hot-pressing	effect on physical properties	BREKHORSKILCH	(224)
α-Al_2O_3	TiO_2	——	mechanical strength	KARPINOS	(225)
α-Al_2O_3	Si_3N_4	——	impact strength	KARPINOS	(226)
BeO	Al_2O_3-BN	——	thermal conductivity and heat resistance	GANTMAN	(227)
BN	MgO	——	strength and heat resistance	KARPINOS	(228)
Cr_2O_3	Cr_2O_3	——	heat resistance	KARPINOS	(229)
MgO	Cr_2O_3	——	abrasion resistance	KARPINOS	(230)
Si_3N_4	ZrO_2	——	heat resistance	KARPINOS	(232)
Si_3N_4	Si_3N_4	sintering	strength improvement	SUMITOMO ELECTRIC IND.	(233)
Si_3N_4	Si_3N_4	sintering	impact strength	KARPINOS	(234)
SiC, BN, C	Si_3N_4, AlN	sintering or hot-pressing	heat resistance	TOKYO SHIBANCRA	(235)
SiO_2	Al_2O_3	——	heat resistance	KARPINOS	(236)
spinel	Cr_2O_3	——	heat resistance	SUVOVROC	(237)
TiO_2	TiO_2	——	heat resistance	LESOVOI	(238)
ZnO	TiO_2	——	impact strength and wear resistance	KARPINOS	(239)
ZrO_2	stab.ZrO_2	hot-pressing	heat resistance and mechanical properties	VOLKOGON	(240)
ZrO_2	MgO	hot-pressing	compressive, bending and impact strength	KARPINOS	(241)
ground whiskers	several oxides	powder metallurgy techniques	heat resistance, compressive strength	INSTITUT VYSOKIKY SSSR	(242)

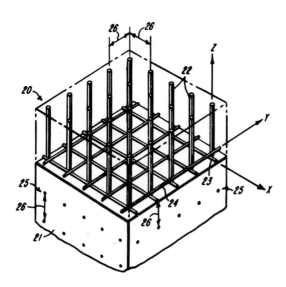

Composite structure reinforced in the X, Y and Z directions and hot pressing apparatus for the production thereof (AVCO CORP)

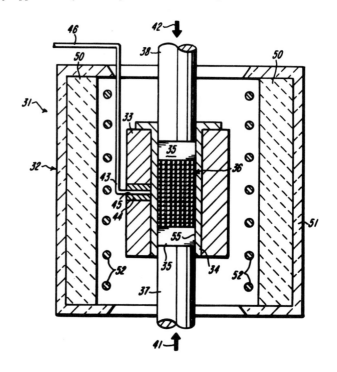

TABLE 8 : Summary of Ceramic Fibre - Ceramic Matrix Composites

FIBRES	MATRIX	FABRICATION METHOD	FIELD OF APPLICATION OR SCOPE OF RESEARCH	AUTHOR	REF.
Al_2O_3	Al_2O_3	sintering	transverse rupture strength	SUMITOMO CHEMICAL CO	(243)
AlN	Al_2O_3	tape casting, aligning AlN needles and sintering	anisotropic thermal properties; substrates for microcircuits	HONEYWELL	(244)
AlN, Si_3N_4	AlN, Si_3N_4	hot-pressing	turbine blades	TOKYO SHIBAURA	(245)
Al_2O_3, C, ZrO_2	$Mg_3(PO_4)_2$	hot-pressing	properties study	KIMURA	(246)
Al_2O_3, C, B, SiC, SiO_2	Al_2O_3, $3Al_2O_3.2SiO_2$	hot-pressing	properties study	FITZER	(247)
Al_2O_3, C, B, BN, SiC	Si_3N_4	reaction sintering	compatibility of fibres and matrix	FISCHBACH	(248)
$3Al_2O_3.2SiO_2$	$3Al_2O_3.2SiO_2$- Al_2O_3	slip casting and firing	biomedical materials	BAER	(249)
BN	Al_2O_3	—	cutting tool	SUWA SEIKOSHA	(250)
BN	BN	—	effect of fibre content on density and bending strength	MOROZOV	(251)
BN	BN	hot-pressing	fabrication process; numerous applics.	CARBORUNDUM	(252)(253)
BN	BN	chemical vapour deposition	hollow BN fibres; aerospace and thermal insulation	SECR. OF STATE OF DEFENCE, LONDON	(254)
BN	BN	firing B_2O_3 contg. composite in nitriding atmosphere	electric cell separator	KENNECOTT	(255)
C	Al_2O_3, $3Al_2O_3.2SiO_2$	coating fibres with LiC; sintering	adhesion improvement	YASUDA	(256)
C	Al_2O_3	hot-pressing	properties study	YOSHIKAWA	(257)(258)
C	Al_2O_3	hot-pressing	gas turbine blades, hot-pressing dies	ASANO	(259)
C	carbides, borides, silicides, oxides	hot-pressing	fabrication process	MC DONNEL DOUGLAS	(260)(261)
C	pyrolytic materials	chemical vapor deposition	aerospace	DUCOMMUN	(262)
C	C-SiC, TiC	chemical vapor deposition	high temperature applics.	SOC. EUR. DE PROPULSION WARREN CONSULTANTS	(263)(264) (265)
C	Si_3N_4	hot-pressing, reaction sintering; coating the fibres e.g. with SiC to improve compatibility	—	PLESSEY CO LUCAS LTD DENKI KAGAKU NAT. RESEARCH	(266) (267)(268) (269) (270)
C	sialon, Si_3N_4, Si_3N_4-C, TaC	hot-pressing	structure reinforced in three mutually perpendicular X, Y and Z directions	AVCO	(45)(46)
C	TaC	vacuum impregnation with precursor solution and pyrolyzing	nose tips for space vehicles	GIBSON	(271)(172)
C	C-TaC	hot-pressing Ta-coated fibres	thermal shock resistance	US DEPARTMENT OF ENERGY	(273)(275)
C	ZrB_2-Si-C	hot-pressing	wetting with eutectic alloy	KRIVOSHEIN	(274)

TABLE 8 (End)

FIBRES	MATRIX	FABRICATION METHOD	FIELD OF APPLICATION OR SCOPE OF RESEARCH	AUTHOR	REF.
C, fused SiO_2	powdered ceramic	applying aqueous slurry and drying	aircraft	GENERAL ELECTRIC	(276)
MgO	Cub.ZrO_2	directional solidification	ZrO_2-MgO	KENNARD	(277)
SiC	Si	impregnation carbon fibre preform with molten silicon	gas turbines	GENERAL ELECTRIC TOSHIBA CERAMICS PAMPUCH	(278 - 281) (282) (283)
SiC	Si	heating a mixture of carbon fibres and silicon powder	gasket	GENERAL ELECTRIC	(284)
SiC	Si	infiltrating silicon carbide fibres with molten silicon	high temperature; resistance to corrosion	RESEARCH INST. FOR IRON, STEEL AND OTHER METALS	(285)
SiC	SiC	chemical vapor deposition	——	FITZER SOC. EUR. DE PROPULSION	(286) (287)
SiC	SiC, Si_3N_4, AlN, BN	hot-pressing or sintering	high temperature	SAMSONOV RESEARCH INST. FOR IRON, STEEL AND OTHER METALS	(288 - 290) (291)
SiC	Si_3N_4	reaction sintering	heat resistance	KUROSAKI REFR. LUCAS LTD NAT. RESEARCH DEVELOPMENT	(292) (293) (270)
Si_3N_4	Si_3N_4	hot-pressing	——	SUMITOMO ELECTRIC	(294)
ZrO_2	Al_2O_3	directional solidification	gas turbine engines	UNITED AIRCRAFT	(295)
ZrO_2	CaO-ZrO_2	directional solidification	gas turbine engines	UNITED AIRCRAFT	(296)
ZrO_2	MgO	hot-pressing	microstructure study	KARPINOS YASUDA	(297) (298)
ZrO_2	ZrO_2	hot-pressing	preparation; mechanical properties	GRAVES	(299)
ZrO_2	ZrO_2	impregnation	heat shields	UNION CARBIDE	(300)

It is remarkable that most work on ceramic whisker reinforced composites has been done in Russia; even so a great deal of the whiskers used are related to mullite whiskers which are used to reinforce oxide and non-oxide ceramics as well as cermets thus showing their chemical and mechanical compatibility with the most diverse matrices.

However fibrous whisker reinforced materials having uniform properties throughout the article produced are difficult to obtain; generally these composites are produced by powder metallurgical techniques: whiskers are difficult to handle and the greatest care must be taken in preparing a homogeneous mixture of these whiskers and matrix powder materials.

In modern advanced technology there is an increasing demand for structural materials having controlled uniform or non-uniform properties e.g. isotropic and anisotropic thermal conductivity. Controlled fabrication processes thus had to be developed. As a result, development of composites containing continuous fibres was started in Europe, the USA and particularly in Japan: these fibres are easy to handle, they can be aligned in one direction, they can be easily stacked in layers to form a laminated structure, the quantity of fibres can be changed from one layer to another, etc., thus enabling the production of materials with predeterminated properties.

As can be seen from the summary given, alumina, carbon and silicon carbide continuous fibres have been mostly used up to now; this is not surprising because high strength fibres of this kind recently became available on the market.

3. Latest Developments and Trends

At the moment the development of fibre reinforced ceramic materials is focused on carbon-carbon composites and continuous fibre reinforced glass on ceramic matrix structural materials.
With respect to carbon-carbon composites, developments in the near future will be probably directed to proposals for the fabrication of multidirectional reinforcement structures in direct relation with specific applications although a great deal of work in the field has been already accomplished.
Developments of continuous fibre reinforced glass and ceramic matrix composites will be strongly dependent on high strength refractory fibre developments: the production of new fibres with improved properties and their commercial availability will lead to testing of their chemical compatibility with various ceramic matrices, especially non-oxide ceramics which are very promising materials in high temperature technology. In the near future attemps to improve the said compatibility of already available fibres will probably include either some pretreatment of the fibres involved or development of suitable matrix compositions. In conclusion it may be assumed that developments of these composite materials have just started and interest in further developments by the industry involved will be growing rapidly.

On the contrary metal fibre-ceramic systems are not very promising for the reasons already mentioned before. Incompatibility of fibres and matrix, and thermal expansion mismatch are problems difficult to solve. Except for special purposes further interesting developments of the metal fibre reinforced ceramics as structural materials are not to be expected.

REFERENCES TO CHAPTER 2

(1)   DE   440745   (R. SCHNABEL)
(2)   H.W. RAUCH, W.H. SUTTON and L.R. McCREIGHT : "Ceramic fibers and
      fibrous composite materials", Academic Press, New York and London,
      1968, p 101 - 102
(3)   T. VASILOS *et al.*, J. of Metals, $\underline{18}$ (5) p 583 - 592 (1966)
(4)   GB   953651   (THE PLESSEY CO)
(5)   GB   918394   (THE PLESSEY CO)
(6)   US 3233985   (WUERTTEM BERGISCHE METALLWARENFABRIK)
(7)   DE 1471152   (FELDMUEHLE A.G.)
(8)   US 3427185   (UNITED AIRCRAFT CORP)
(9)   D.W. LEVINSON, Trans. Brit. Cer. Soc., $\underline{63}$ (1) 21A (1964)
(10) US 3321285   (MINNESOTA MINING & MANUFACTURING CO)
(11) GB 1151464   (TRW INC)
(12) FR 1579111   (NORTH AMERICAN ROCKWELL CORP)
(13) GB   919181   (THE GENERAL ELECTRIC CO)
(14) GB   954285   (T.I. GROUP SERVICES LTD)
(15) GB 1108633   (ASSOCIATED ELECTRICAL INDUSTRIES LTD)
(16) US 3386918   (USA SECRETARY OF THE AIR FORCE)
(17) US 3384578   (USA SECRETARY OF THE AIR FORCE)
(18) US 3386840   (MONSANTO CO)
(19) US 3462289   (THE CARBORUNDUM CO)
(20) US 3416944   (USA SECRETARY OF THE AIR FORCE)
(21) FR 1493696   (SOC. LE CARBONE-LORRAINE)
(22) GB 1163979   (DUCOMMUN INCORP)
(23) M. CLAUSSEN, J. Am. Cer. Soc., $\underline{56}$ (8) 442 (1973)
(24) JP 7355105   (TOYOTA MOTOR CO)
(25) A.I. AKIMENKO *et al.*, Chem. Abstr., $\underline{81}$ (26) p 360 - 175.1082 (1974)
(26) N. CLAUSSEN *et al.*, Composites, $\underline{6}$ (2) 86 (1975)
(27) US 4103063   (UNITED TECHNOLOGIES CORP)
(28) SU   833872   (A.V. SHEVCHENKO)
(29) M. CLAUSSEN *et al.*, J. Am. Cer. Soc., $\underline{59}$ (B/4) 182 (1976)
(30) US 3706583   (J.C. FLETCHER)
(31) SU   492506   (D.M. KARPINOS)
(32) P. HING *et al.*, J. Mat. Sci., $\underline{7}$ (4) p 427 - 434 (1972)
(33) C.-C. JEN *et al.*, J. Am. Cer. Soc., $\underline{57}$ (5) p 232 - 233 (1974)
(34) J.G. ZWISSLER *et al.*, J. Am. Cer. Soc., $\underline{60}$ (9/10) p 390 - 396 (1977)
(35) US 3794551   (USA SECRETARY OF THE ARMY)
(36) R.H. DUNGAN *et al.*, J. Am. Cer. Soc., $\underline{56}$ (6) 345 (1973)
(37) B.V. LUKIN *et al.*, Chem. Abstr., $\underline{85}$ (22) 371 - 165.539a (1976)
(38) J.J. BRENNAN *et al.*, AD - A051657 (1977)
(39) J.J. BRENNAN *et al.*, AD - A025901 (1976)
(40) J.J. BRENNAN *et al.*, J. Adhes., $\underline{5}$ (2) p 139 - 159 (1973)
(41) J.J. BRENNAN *et al.*, AD - 757.063 (1973)
(42) US 3914500   (UNITED AIRCRAFT CORP)

(43)   US 3900626   (UNITED AIRCRAFT CORP)
(44)   GB 1198906   (NATIONAL RESEARCH DEVELOPMENT CORP)
(45)   US 4165355   (AVCO CORP)
(46)   US 4209560   (AVCO CORP)
(47)   US 3779716   (USAEC)
(48)   US 4067742   (J.C. FLETCHER)
(49)   DE 2113129   (E. FITZER)
(50)   US 3922411   (AVCO CORP)
(51)   CH 586166   (LES FABRIQUES D'ASSORTIMENTS REUNIES)
(52)   DE 2711219   (H. SCHEICHER)
(53)   E. FITZER *et al.*, Sprechsaal, <u>113</u> (12) p 919 - 928 (1980)
(54)   E. FITZER *et al.*, Carbon, <u>18</u> (6) p 383 - 387 (1980)
(55)   E. FITZER *et al.*, Carbon, <u>18</u> (4) p 291 - 295 (1980)
(56)   V. MARKOVIC *et al.*, Carbon, <u>18</u> (5) p 329 - 335 (1980)
(57)   E. FITZER *et al.*, Carbon, <u>18</u> (4) p 265 - 270 (1980)
(58)   W.G. BRADSHAW *et al.*, Am. Cer. Soc. Bull., <u>57</u> (2) p 193 - 198 (1978)
(59)   T.R. GUESS *et al.*, Composites, <u>7</u> (3) 200 (1976)
(60)   EP   57637   (SOCIETE EUROPEENNE DE PROPULSION)
(61)   EP   32858   (SOCIETE EUROPEENNE DE PROPULSION)
(62)   FR 2444012   (SOCIETE EUROPEENNE DE PROPULSION)
(63)   FR 2427198   (SOCIETE EUROPEENNE DE PROPULSION)
(64)   FR 2424888   (SOCIETE EUROPEENNE DE PROPULSION)
(65)   FR 2421056   (SOCIETE EUROPEENNE DE PROPULSION)
(66)   FR 2276916   (SOCIETE EUROPEENNE DE PROPULSION)
(67)   EP   37104   (EATON CORP)
(68)   EP   27251   (FORDATH LTD)
(69)   EP   29851   (THE B.F. GOODRICH CO)
(70)   FR 2376831   (C. CONRADTY NUERNBERG GmbH)
(71)   FR 2296592   (COAL INDUSTRY LTD)
(72)   US 4048953   (PFIZER INC)
(73)   FR 2144330   (THE GOODYEAR TIRE & RUBBER CO)
(74)   FR 2111672   (COAL INDUSTRY LTD)
(75)   FR 2488244   (TOHO BESLON CO)
(76)   FR 2446175   (SOCIETE EUROPEENNE DE PROPULSION)
(77)   FR 2434964   (V.V. KULAKOV)
(78)   FR 2433003   (COMMISSARIAT A L'ENERGIE ATOMIQUE)
(79)   FR 2427315   (COMMISSARIAT A L'ENERGIE ATOMIQUE)
(80)   FR 2427197   (SOCIETE EUROPEENNE DE PROPULSION)
(81)   FR 2416391   (GOODYEAR AEROSPACE CORP)
(82)   FR 2414574   (DUNLOP LTD)
(83)   FR 2404524   (THE CARBORUNDUM CO)
(84)   FR 2402631   (KANEBO LTD)
(85)   FR 2394507   (KANEBO LTD)
(86)   FR 2398705   (LE CARBONE-LORRAINE)
(87)   FR 2391956   (C. CONRADTY NUERNBERG GmbH)
(88)   FR 2386890   (EURATOM)
(89)   FR 2378888   (MORGANITE MODMOR LTD)
(90)   FR 2369230   (N.V. PHILIPS GLOEILAMPENFABRIEKEN)
(91)   FR 2334495   (SOCIETE EUROPEENNE DE PROPULSION)
(92)   FR 2313601   (GOODYEAR AEROSPACE CORP)
(93)   FR 2289459   (DUCOMMON INCORP)
(94)   FR 2276913   (SOCIETE EUROPEENNE DE PROPULSION)
(95)   FR 2270203   (DUNLOP LTD)
(96)   FR 2260726   (DUNLOP LTD)
(97)   FR 2225654   (GOODYEAR TIRE & RUBBER CO)
(98)   FR 2196966   (HYFIL LTD)
(99)   FR 2190728   (KUREHA KAGAKU KOGYO K.K.)

(100) US 3991248 (DUCOMMUN INCORP)
(101) FR 2187417 (BRITISH NUCLEAR FUELS LTD)
(102) FR 2171414 (SIGRI ELEKTROGRAPHIT GmbH)
(103) FR 2144329 (THE GOODYEAR TIRE & RUBBER CO)
(104) FR 2143124 (THE BENDIX CORP)
(105) US 3867491 (THE CARBORUNDUM CO)
(106) US 3759353 (THE CARBORUNDUM CO)
(107) US 3712428 (THE CARBORUNDUM CO)
(108) FR 2091204 (UKAEA)
(109) US 3943213 (GREAT LAKES CARBON)
(110) FR 2087202 (COMMISSARIAT A L'ENERGIE ATOMIQUE)
(111) FR 2086156 (KUREHA KAGAKU KOGYO K.K.)
(112) FR 2065763 (LE CARBONE-LORRAINE)
(113) FR 1594182 (THE CARBORUNDUM CO)
(114) DE 2912786 (AUDI NSU AUTO UNION)
(115) DE 2714364 (SCHUNK & EBE GmbH)
(116) DE 2659374 (NIPPON CARBON CO)
(117) DE 2623968 (NIPPON CARBON CO)
(118) DE 2430719 (D. BITZER)
(119) DE 2165029 (KUREHA KAGAKU KOGYO K.K.)
(120) DE 1930713 (KUREHA KAGAKU KOGYO K.K.)
(121) DE 2103908 (E. FITZER)
(122) DE 1961303 (USAEC)
(123) US 3671385 (USAEC)
(124) DE 1928373 (BROWN BOVERI)
(125) DE 1571320 (R.L. BICKERDIKE)
(126) GB 2099365 (LE CARBONE-LORRAINE)
(127) GB 2053177 (UNION CARBIDE CORP)
(128) GB 1549687 (DUNLOP LTD)
(129) GB 1546802 (SHELL INTERNATIONALE RESEARCH MAATSCHAPPIJ)
(130) GB 1475306 (SHELL INTERNATIONALE RESEARCH MAATSCHAPPIJ)
(131) GB 1469754 (DUNLOP LTD)
(132) GB 1457757 (SECRETARY OF STATE OF DEFENCE, LONDON)
(133) GB 1455331 (UKAEA)
(134) GB 1343773 (UKAEA)
(135) GB 1434824 (COAL INDUSTRY LTD)
(136) GB 1421672 (UKAEA)
(137) GB 1410090 (UKAEA)
(138) GB 1352141 (NATIONAL RESEARCH DEVELOPMENT CORP)
(139) GB 1330970 (FORDATH LTD)
(140) GB 1230970 (FORDATH LTD)
(141) GB 1312258 (THE PLESSEY CO)
(142) US 3720575 (THE PLESSEY CO)
(143) GB 1236015 (COURTAULDS LTD)
(144) US 4252588 (SCIENCE APPLICATIONS)
(145) US 4201611 (USA - SECRETARY OF THE AIR FORCE)
(146) US 4193828 (FIBER MATERIALS)
(147) US 4131708 (FIBER MATERIALS)
(148) US 4166145 (HITCO)
(149) US 4026745 (HITCO)
(150) US 4164601 (HITCO)
(151) US 4104354 (HITCO)
(152) US 4152482 (USA - US DEPARTMENT OF ENERGY)
(153) US 3793204 (USA - US DEPARTMENT OF ENERGY)
(154) US 4152381 (USA - US DEPARTMENT OF ENERGY)
(155) US 4123832 (GENERAL ELECTRIC CO)
(156) US 4064207 (UNITED TECHNOLOGIES)

(157) US 4041116   (GREAT LAKES CARBON CORP)
(158) US 4029844   (ATLANTIC RESEARCH CORP)
(159) US 3935354   (ATLANTIC RESEARCH CORP)
(160) US 3925133   (ATLANTIC RESEARCH CORP)
(161) US 3900675   (ATLANTIC RESEARCH CORP)
(162) US 3897582   (ATLANTIC RESEARCH CORP)
(163) US 3717419   (ATLANTIC RESEARCH CORP)
(164) US 3949126   (FIBER MATERIALS)
(165) US 3936552   (THE B.F. GOODRICH CO)
(166) US 3927157   (HITCO)
(167) US 3917884   (FIBER MATERIALS)
(168) US 3914395   (AVCO CORP)
(169) US 3856593   (HAVEG INDUSTRIES)
(170) US 3796616   (HAVEG INDUSTRIES)
(171) US 3758352   (HAVEG INDUSTRIES)
(172) US 3772115   (THE CARBORUNDUM CO)
(173) US 3728423   (HITCO)
(174) US 3718720   (USAEC)
(175) US 3700535   (USAEC)
(176) US 3550213   (USA - SECRETARY OF THE ARMY)
(177) FR 2014130   (UKAEA)
(178) FR 2027513   (ROLLS-ROYCE LTD)
(179) US 3575789   (OWENS-CORNING FIBERGLASS CORP)
(180) US 3607608   (OWENS-CORNING FIBERGLASS CORP)
(181) US 3792985   (OWENS-CORNING FIBERGLASS CORP)
(182) GB 1403863   (UKAEA)
(183) D.C. PHILLIPS, J. Mat. Sci. 9 (11), p 1847 - 1854 (1974)
(184) S.R. LEVITT, J. Mat. Sci. 8 (6), p 793 - 806 (1973)
(185) R.A.J. SAMBELL et al., Chem. Abstr. 84 (14), 288 - 93.344Z (1976)
(186) K.R. LINGER et al., Composites 8 (3), 139 - 144 (1977)
(187) E. FITZER et al., Rev. Int. Hautes Temp. Réfract. 16 (2),
       p 147 - 155 (1979)
(188) M. SAHEBKAR et al., Ber. Dtsch. Keram. Ges. 55 (5), p 265 - 268 (1978)
(189) K.M. PREWO et al., Chem. Abstr. 91 (6), 250 - 43.432p (1979)
(190) K.M. PREWO et al., Chem. Abstr. 95 (6),283 - 47.652r (1981)
(191) K.M. PREWO et al., Chem. Abstr. 94 (10),289 - 70.178d (1981)
(192) K.M. PREWO et al., Chem. Abstr. 94 (10),288 - 70.173y (1981)
(193) US 4263367   (UNITED TECHNOLOGIES CORP)
(194) US 4265968   (UNITED TECHNOLOGIES CORP)
(195) J.F. BACON, Chem. Abstr. 88 (12), 235 - 77.799s (1978)
(196) J.F. BACON et al, Chem. Abstr. 94 (10), 288 - 70.174z (1981)
(197) US 4268562   (UNITED TECHNOLOGIES CORP)
(198) K.M. PREWO et al., J. Mat. Sci. 15 (2), p 463 - 468 (1980)
(199) FR 2475534   (UNITED TECHNOLOGIES CORP)
(200) GB 2075490   (UNITED TECHNOLOGIES CORP)
(201) US 4341826   (UNITED TECHNOLOGIES CORP)
(202) D.M. KARPINOS et al., Chem. Abstr. 96 (2),265 - 10.664r (1982)
(203) E.R. THOMPSON et al., Chem. Abstr. 95 (16),271 - 137.050x (1981)
(204) US 3828699   (UKAEA)
(205) US 3901719   (CORNING GLASS WORKS)
(206) US 3948669   (CORNING GLASS WORKS)
(207) J.P. LUCAS et al., J. Am. Cer. Soc. 63 (5), p 280 - 285 (1980)
(208) I.W. DONALD et al., J. Mat. Sci. 12 (2),p 290 - 298 (1978)
(209) DE 3018465   (THE RESEARCH INSTITUTE FOR SPECIAL INORGANIC MATERIALS)
(210) D.M. KARPINOS et al., Glass and Ceramics 39 (1/2),p 84 - 85 (1982)
(211) D.M. KARPINOS et al., Chem. Abstr. 91 (4), 334 - 25.987u (1979)
(212) D.M. KARPINOS et al., Chem. Abstr. 92 (10),292 - 81.009j (1980)

(213) D.M. KARPINOS *et al.*, Glass and Ceramics 35 (5/6), p 355 - 357 (1978)
(214) SU 395342   (D.M. KARPINOS)
(215) G.A. TOTSKAYA, Chem. Abstr. 87 (16), 258 - 121.740w (1977)
(216) E.L. GUMENYUK, Chem. Abstr. 87 (1), 294 - 89.341d (1977)
(217) SU 392046   (D.M. KARPINOS)
(218) SU 422705   (G.V. SAMSONOV)
(219) SU 415247   (D.M. KARPINOS)
(220) SU 393251   (D.M. KARPINOS)
(221) D.M. KARPINOS *et al.*, Glass and Ceramics 31 (5/6), p 345 - 346 (1974)
(222) D.M. KARPINOS *et al.*, Glass and Ceramics 31 (3/4), p 200 - 202 (1974)
(223) SU 386874   (G.V. SAMSONOV)
(224) S.M. BREKHOVSKIKH *et al.*, Glass and Ceramics 36 (5/6), p 266 - 269 (1979)
(225) SU 478818   (D.M. KARPINOS)
(226) SU 390049   (D.M. KARPINOS)
(227) SU 833850   (S.A. GANTMAN)
(228) SU 553228   (D.M. KARPINOS)
(229) SU 414232   (D.M. KARPINOS)
(230) SU 477974   (D.M. KARPINOS)
(231) SU 381647   (D.M. KARPINOS)
(232) SU 381650   (D.M. KARPINOS)
(233) JP 8192180   (SUMITOMO ELECTRIC IND)
(234) SU 380615   (D.M. KARPINOS)
(235) US 3833389   (TOKYO SHIBAURA ELECTRIC CO)
(236) SU 381645   (D.M. KARPINOS)
(237) SU 528286   (S.A. SUVOROV)
(238) SU 458535   (M.V. LESOVOI)
(239) SU 483378   (D.M. KARPINOS)
(240) L.M. VOLKOGON *et al.*, Chem. Abstr. 81 (18), 320 - 110.372c (1974)
(241) D.M. KARPINOS *et al.*, Chem. Abstr. 90 (22), 305 - 173.323u (1979)
(242) GB 1448918   (INSTITUT VYSOKIKH TEMPERATUR AKADEMII NAUK-SSSR)
(243) JP 7747803   (SUMITOMO CHEMICAL CO)
(244) US 4256792   (HONEYWELL INC)
(245) JP 7912488   (TOKYO SHIBAURA ELECTRIC CO)
(246) S. KIMURA *et al.*, Chem. Abstr. 95 (22), 266 - 191.298x (1981)
(247) E. FITZER *et al.*, High Temp. Sci. 13 (1 - 4), p 149 - 172 (1980)
(248) D.B. FISCHBACH, Chem. Abstr. 92 (14), 275 - 115.218u (1980)
(249) J.R. BAER *et al.*, Am. Cer. Soc. Bull. 57 (2), p 220 - 222 (1978)
(250) JP 7440121   (SVWA SEIKOSHA CO)
(251) S.V. MOROZOV *et al.*, Chem. Abstr. 95 (1), 267 - 11.387u (1981)
(252) R.Y. LIN *et al.*, Am. Cer. Soc. Bull. 55 (9), p 781 - 784 (1976)
(253) US 4075276   (THE CARBORUNDUM CO)
(254) GB 2014972   (THE SECRETARY OF STATE OF DEFENCE, LONDON)
(255) US 4284610   (KENNECOTT CORP)
(256) E. YASUDA *et al.*, Z. Werkstofftech. 9 (9), p 310 - 315 (1978)
(257) M. YOSHIKAWA *et al.*, Chem. Abstr. 89 (6), 249 - 47.856m (1978)
(258) M. YOSHIKAWA *et al.*, Chem. Abstr. 86 (6), 234 - 33.347x (1977)
(259) JP 75136306 (T. ASANO)
(260) US 3766000   (McDONNELL DOUGLAS CORP)
(261) US 3736159   (McDONNELL DOUGLAS CORP)
(262) US 3991248   (DUCOMMUN INC)
(263) FR 2401888   (SOCIETE EUROPEENNE DE PROPULSION)
(264) R. MASLAIN *et al.*, Rev. Chim. Miner. 18 (5), p 544 - 564 (1981)
(265) US 4275095   (WARREN CONSULTANTS)
(266) GB 1264476   (THE PLESSEY CO LTD)
(267) FR 2053222   (JOSEPH LUCAS LTD)
(268) FR 2053221   (JOSEPH LUCAS LTD)

(269)   JP 7732906   (DENKI KAGAKU KOGYO K.K.)
(270)   FR 2011863   (NATIONAL RESEARCH DEVELOPMENT CORP)
(271)   US 4196230   (J.O. GIBSON)
(272)   US 4278729   (J.O. GIBSON)
(273)   US 4180428   (US DEPARTMENT OF ENERGY)
(274)   D.A. KRIVOSHEIN, Chem. Abstr. $\underline{93}$ (8), 398 - 78.181Z (1980)
(275)   L.R. NEWKIRK *et al.*, Proc.-Electrochem. Soc., 79-3, p 488 - 498
        (1979)
(276)   US 4284664   (GENERAL ELECTRIC CO)
(277)   F.L. KENNARD *et al.*, J. Am. Cer. Soc. $\underline{57}$ (10), p 428 - 431 (1974)
(278)   FR 2444650   (GENERAL ELECTRIC CO)
(279)   US 4294788   (GENERAL ELECTRIC CO)
(280)   US 4240835   (GENERAL ELECTRIC CO)
(281)   US 4238433   (GENERAL ELECTRIC CO)
(282)   JP 7990209   (TOSHIBA CERAMICS CO)
(283)   R. RAMPUCH *et al.*, Chem. Abstr. $\underline{96}$ (10) 298 - 73.490b (1982)
(284)   FR 2373348   (GENERAL ELECTRIC CO)
(285)   FR 2347463   (THE RESEARCH INSTITUTE FOR IRON, STEEL AND OTHER
                      METALS OF THE TOHOKU UNIVERSITY)
(286)   E. FITZER, Chem. Abstr. $\underline{93}$ (24), 250 - 224.595f (1980)
(287)   EP   32097   (SOCIETE EUROPEENNE DE PROPULSION)
(288)   SU 374256   (G.V. SAMSONOV)
(289)   SU 374258   (G.V. SAMSONOV)
(290)   SU 356264   (G.V. SAMSONOV)
(291)   FR 2329611   (THE RESEARCH INSTITUTE FOR IRON, STEEL AND OTHER
                      METALS OF THE TOHOKU UNIVERSITY)
(292)   JP 81129668   (KUROSAKI REFRACTORIES CO)
(293)   FR 2015951   (JOSEPH LUCAS LTD)
(294)   JP 81100168   (SUMITOMO ELECTRIC IND)
(295)   GB 1401371   (UNITED AIRCRAFT CORP)
(296)   US 3887384   (UNITED AIRCRAFT CORP)
(297)   D.M. KARPINOS *et al.*, Chem. Abstr. $\underline{95}$ (24), 280 - 208.276p (1981)
(298)   E. YASUDA *et al.*, Chem. Abstr. $\underline{86}$ (1), 324 - 77.539r (1977)
(299)   G.A. GRAVES *et al.*, Am. Cer. Soc. Bull. $\underline{49}$ (9), 797 - 803 (1970)
(300)   GB 1353384   (UNION CARBIDE CORP)

CHAPTER 3

General Methods for the Manufacture of Composite Materials

1. Introduction

The fabrication methods which have been used up to now, consist in fact of a combination of technical operations each of which has been well known perse for a long time. Hence development of these methods generally implies the finding of a more optimalized combination of steps which can reduce costs or which is better adapted to the manufacture of new composites.
Making a distinction between the state of the art prior to 1970 and the more recent technique therefore seemed meaningless and consequently the present chapter will give an overall survey of all basic methods which have been reported up to now.

2. Methods based on the Combination of Preformed Fibres with Matrix Material

Such methods usually involve three basic steps:

(1) Alignment of the fibres or filaments into a specific configuration or pattern.
For longer fibres or filaments this can be achieved by prearrangement by winding, formation of three dimensional structures, etc.
Shorter fibres such as whiskers, are advantageously aligned by magnetic, electrostatic, vibratory, rolling or flotation techniques. Alignment can also be performed during the stage where the matrix is combined with the fibres and subsequently shaped e.g. by extrusion, drawing, ...

(2) Combination of the matrix material with the fibres.
Here numerous techniques can be used depending on whether the matrix material will be applied as a solid or liquid or will be deposited molecularly. Powder metallurgical techniques, liquid impregnation and vapor deposition are typical examples.

(3) Consolidation of the combined constituents.
This is normally obtained by compression and/or heat treatment, a multitude of variants again being possible, such as isostatic hot pressing, liquid phase hot pressing, etc.

It is obvious that the selection of a given process combination will in the first place be dictated by parameters such as the physical or chemical properties of the fibres and the matrix material (e.g. melting point, brittleness,...) and further by the desired mechanical properties and final shape of the composite. But in any case the selected combination should be compatible with the following criteria :

123

- permit the desired distribution and alignment of the fibres in the finished composite;
- develop a good bond between matrix and fibres;
- avoid deterioration of the strength properties of the fibres by breakage, surface reaction, etc.
- allow economically feasible production.

The basis processes which have been reported up to now are summarized in the following three tables pertaining respectively to techniques whereby the matrix material is originally present as a solid, or liquid or is deposited in the molecular state.

TABLE 9: Solid State Matrix Processes

ORIGINAL FORM OF THE MATRIX MATERIAL	COMBINATION TECHNIQUE	CONSOLIDATION TECHNIQUE	FIELD OF APPLICATION	REF
Preformed elements; sheets, foils, etc.	stacking of layers of fibres and sheets	rolling, sintering, diffusion bonding or	metal-metal com- posites	(1)(2)(3) (4)(5)
	stacking of layers of fibres and sheets	heating under pressure to melt the matrix	metal-metal com- posites	(6)(7)
tubes	inserting reinforcing fibres in matrix tubes	drawing, extrusion, diffusion bonding	metal matrix composites	(8)(9)
fibres, strips, etc.	cowinding reinforcing fibres and matrix strips	extrusion, pressing; heat treatment for diffusion bonding or to melt the matrix	metal matrix composites	(10)(11)
Powder; dry powder	mixing matrix with fibres	(isostatic) pressing, hot pressing	various	(12)(13) (14)(15) (21)
	mixing matrix with fibres	liquid phase sintering	various	
	mixing matrix with fibres	explosion compression	various	(16)(17)
	mixing matrix with fibres	continuous by rolling, laminating	various	(18)
	mixing matrix with fibres	extrusion	various	(19)(20)
	infiltrating powder into reinforcing structure	compression, sintering	various	(27)(28)
suspension, slurry	casting of matrix	sintering, isostatic pressing	ceramic matrix composites	(22)(23) (24)(25)
	dipping of fibres in matrix slurry	sintering, isostatic pressing		(26)

TABLE 10: Liquid Matrix Processes

ORIGINAL FORM OF THE MATRIX MATERIAL	COMBINATION TECHNIQUE	CONSOLIDATION TECHNIQUE	FIELD OF APPLICATION	REF
melts	impregnation of fibre structure, possibly under pressure	solidification of the matrix	light metal matrix composites	(29)(30)(33)
	casting melt containing fibres	solidification of the matrix	whisker reinforced metal matrix composites	(31)
	extrusion of melt containing fibres	solidification of the matrix	whisker reinforced metal matrix composites	(32)
	passing fibres through melt	solidification of the matrix	metal composites reinforced with filaments	(34)(35)(36)(37)(38)
molten particles	plasma spraying	solidification of the matrix	composites with matrix of refractory metals or refractory ceramics	(39)(40)(41)
melts or solutions of matrix precursor material	impregnation, dipping, spraying	curing and hardening of matrix thermal decomposition	carbon-carbon composites	(42)

TABLE 11: Molecular Deposition Processes

DEPOSITION METHOD	CONSOLIDATION	FIELD OF APPLICATION	REF
electrolytic	as deposited or thermal or thermo-mechanical after-treatment	metal matrix composites in part. Ni	(43)(44)(45)
electrophoretic	compression and sintering	ceramic matrix composites	(46)
codeposition of fibres and matrix	compression and sintering	whisker reinforced composites	(47)(48)(49)
chemical vapor deposition	compression and sintering	composites with carbon, carbide, boride, nitride matrix	(50)(51)(52)(53)(54)

3. Methods involving the *in situ* Generation of Fibres

3.1. Mechanical Processes.   Solid particles of materials with sufficient
ductility can be fibered when subjected to shearing forces. This principle
has been applied to produce composites containing a fibrous like strenthe-
ning phase by mechanical working by drawing, swaging or extrusion of a
powder mixture containing particles which at the working temperature have
sufficient ductility.
Said method has particularly been used for metals which are difficult to
process e.g. refractory metals and has proved to be economically attractive
as it does not need separate preparation of the fibres (144)(145).

In a somewhat different embodiment, the two phases of the composite are
molten and the mixture is extruded resulting in the formation of fine fibres
in a solidified matrix (146).

3.2. Unidirectional Solidification of Eutectic Melts.   Solidification of a
binary eutectic melt involves the formation of two solid phases in equilibrium
with the eutectic melt.
By unidirectionally solidifying such a melt, both solid phases may grow
simultaneously in a parallel array. One phase may have a rodlike structure
or a lamellar structure, bonded into a matrix forming second phase.
Such an unidirectionally solidifying process produces, in one step, directly
from the melt a finished composite formed *in situ* and having a uniform
distribution of aligned, well bonded monocrystalline fibres or lamellae in a
matrix. This eliminates in one single step the costly and slow solutions of
the conventional composite making processes for handling fibres and for
incorporating them with well bonded interfaces into a matrix material.

Several phenomena remain unexplained in understanding the causes of lamellar
or fibrelike growth; some eutectics exhibiting both structures.
Since both structures result in similar properties, both structures are
treated in this review.

Systems of more than two elements may also contain eutectics, *i.e.*points of
fixed temperature and composition at which more than two phases,
simultaneously solidify from the liquid. Also they may be solidified along a
so-called (pseudo) eutectic line exhibiting a eutectic-type behaviour, in
which case the system is called monovariant, bivariant, etc.

Directional solidification can be accomplished by most of the techniques
usually used for single crystal growth from the melt.
Normal solidification (the BRIDGMAN technique) has been used most commonly.
For higher temperature or more reactive materials the floating-zone-tech-
nique is used. Also the crystal pulling technique (the CZOCHRALSKI technique)
and the EFG-technique (edge-defined, film-fed growth) have been introduced.

Directionally-solidified eutectic composites are most commonly used as high-
strength superalloys, for applications such as turbine blades and vanes in
the high-temperature sections of advanced jet engines. Two large jet-engine
manufacturers as UNITED AIRCRAFT, PRATT & WHITNEY and GENERAL ELECTRIC, have
been demonstrating considerable activity in this field.

Extensive effort has been expended on unidirectionally solidified eutectics
for optical, electronic and magnetic applications. The most recent development in this field is the fabrication of optical fibres.

The materials grown by the eutectic solidification technique are summarized
in the form of the following table.

TABLE 12 : Data for Composite Materials.

MICROSTRUCTURE	MATRIX	SOLIDIFICATION TECHNIQUE	PROPERTIES OR APPLICATION FIELD	AUTHOR(S)	REF.
Ti_5Si_3 fibres	Ti base solution	zone melting	mechanical strength	PRUD'HOMME	(55)
NiBe fibres	Ni-Cr solution	normal solidification	mechanical strength	SHEN	(56)
$CuCd_3$ rods Mg_2Si rods	Cd Mg	normal solidification	—	HAOUR	(57)
ZrCuSi fibres	Cu	zone melting	mechanical strength	SPRENGER	(58)
Mo fibres	NiAlTa alloy	normal solidification	mechanical strength	PEARSON	(59)
γ (Ni, Al, Ta) composition fibres (structure based on Ni_3Al)	γ Ni base solution	normal solidification	mechanical strength	JACKSON	(60)
β (Ni, Fe, Al) composition lamellae	γ Ni base solution	normal solidification	mechanical strength	JACKSON	(61)
β (Ni, Co) Al lamellae	γ Ni base solution	normal solidification	mechanical strength	JACKSON	(62)
Mo fibres	Ni_3Al	normal solidification	mechanical strength	LEMKEY	(63)
Co_2Si fibres or lamellae	CoSiX alloy X = Al, Ga or both	normal solidification	mechanical strength	LIVINGSTON	(64)
Co_2Si irregular fibres or lamellae	CoSiX alloy X = Ta, Nb, V	normal solidification	mechanical strength	LIVINGSTON	(65)
Cr-rich lamellae	Ni-rich alloy	different techniques	mechanical strength	SHAW	(66)
Ni_3Al fibres	Ni alloy	normal solidification	mechanical strength	UNITED TECHNOLOGIES	(67)
Mo alloy fibres	γ-Ni base alloy	normal solidification	mechanical strength	UNITED TECHNOLOGIES	(68)
Mo_2NiB_2 fibres	Ni base alloy	normal solidification	high-temperature strength	SPRENGER	(69)
Al_6Fe fibres	Al base alloy	normal solidification	mechanical strength	COMALCO ALUMINIUM	(70)
Co fibres	CoAl alloy	normal solidification	mechanical strength	HUBERT	(71)
Co fibres (with fibre to lamella transition)	CoAlNi alloy	normal solidification	mechanical strength	HUBERT	(72)
Cr fibres Mo fibres	NiAl alloy NiAl alloy	normal solidification	—	WALTER	(73)
Ni base lamellae ε-Ni_2In rods	ε-Ni_2In β-NiIn	normal solidification	—	LIVINGSTON	(74)
$CrSi_2$ fibres	Si	—	—	LEVINSON	(75)
Al_3Ni fibres	Al	normal solidification	mechanical strength	GRABEL	(76)
Cr fibres	NiAl	normal solidification	mechanical strength	WALTER	(77)
ε-FeSi fibres or lamellae	x-Fe_2Si_5	normal solidification	—	NISHADA	(78)

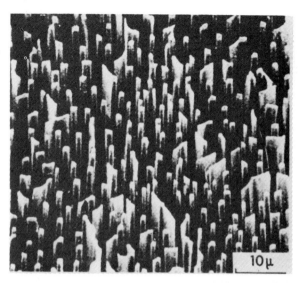

*Microstructure of an unidirectional solidified eutectic
Co-Cr-Ni-TaC alloy, after selective etching of the matrix*

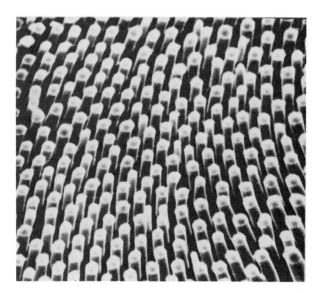

*Scanning electron micrograph of the eutectic alloy
at a cell center showing the morphology of the
intermetallic fibers. Magnification 1882 times.*

TABLE 12 (Cont.)

MICROSTRUCTURE	MATRIX	SOLIDIFICATION TECHNIQUE	PROPERTIES OR APPLICATION FIELD	AUTHOR(S)	REF.
Cr(Mo) plates	NiAl	normal solidification	high temperature strength	WALTER	(79)
Ni_3Al fibres	Ni_3Ta	normal solidification	high temperature strength	ETAT FRANCAIS	(80)
Al_3Ni fibres	Al	EFG	—	TYCO	(81)
Fibres of M, W, V, Ta, Nb, Re or Cr	Cr_2O_3, $LaCrO_3$, ZrO_2, HfO_2, CeO_2, UO_2, ThO_2, Y_2O_3, Al_2O_3, Gd_2O_3 and some mixtures thereof	zone melting	—	U.S. DEPARTMENT OF ENERGY	(82)
M_7C_3 fibres M = Cr, or Co	Co, Ni or Fe base alloy	normal solidification	mechanical strength	UNITED AIRCRAFT	(83)
$(Cr,Co)_{23}C_6$	CoCr base alloy	normal solidification	mechanical strength	UNITED AIRCRAFT	(84)
TaC fibres	Co-Cr base alloy	normal solidification	mechanical strength	WALTER	(85)
TaC fibres	CoCr or NiCr base alloy	normal solidification	mechanical strength	BIBRING	(86)
NbC,TiC,HfC fibres (or solid solutions thereof)	Co or Ni or Fe alloy	normal solidification	mechanical strength	BIBRING	(86)
Cr_2C_2 fibres	Ni alloy	normal solidification	mechanical strength	SPILLER	(87)
TaC fibres	Co, Cr, Ni solution	normal solidification	mechanical strength	DUNLEVEY	(88)
TaC fibres	Co-Cr	normal solidification	mechanical strength	DUNLEVEY	(88)
TaC fibres	two-phase γ-γ' Ni-Cr-Al-Ti alloy;	normal solidification	mechanical strength	BUCHANAN	(89)(90)
TaC fibres	γ phase Ni-Cr alloy				
TaC, VC fibres (or mixture thereof)	Ni base or Co base	normal solidification	mechanical strength	MOORE	(91)
TaC	CoNiWCrTa alloy	normal solidification	mechanical strength	WALTER	(92)
$(Cr,Fe)_7Cr_3$ needles	α and/or γ Fe base alloy	different techniques	mechanical strength	VAN DEN BOOMGAARD	(93)
(W,X)C lamellae C = IVa metal	W base alloy	casting technique	wear resistance	RUAY	(94)
TaC needles	Ta	zone melting	mechanical strength	LEMKEY	(95)
Carbides of Ti,V, Cb,Hf,Zr or Ta or mixtures thereof: fibres	Ni base alloy	normal solidification	mechanical strength	GIGLIOTTI	(96)
TaC	Re containing Ni base alloy	normal solidification	mechanical strength	HENRY	(97)
Cr_3C_2	Ni_3Al	normal solidification	mechanical strength	CHADWICK	(98)
Nb_2C Ta_2C	Nb Ta	different techniques	mechanical strength	NATIONAL RESEARCH DEVELOPMENT	(99)
NbC	NiCoAl base alloy	normal solidification	mechanical strength	O.N.E.R.A.	(100)

TABLE 12 (Cont.)

MICROSTRUCTURE	MATRIX	SOLIDIFICATION TECHNIQUE	PROPERTIES OR APPLICATION FIELD	AUTHOR(S)	REF.
$(Cr,Co,Ni)_7C_3$ fibres	CoCr base alloy	normal solidification	mechanical strength	BROWN, BOVERI	(101)
Fibres of transition metal carbides	Co and Ni base alloy	normal solidification	mechanical strength	O.N.E.R.A.	(102)
Fibres of TaC,VC,WC or mixtures thereof	Ni or Co base alloy	normal solidification	mechanical strength	GENERAL ELECTRIC	(103)
M_7C_3 fibres M = Fe, Co or Cr	FeCrCo alloy	normal solidification niques	mechanical strength	PHILIPS	(104)
TaC fibres	NiCrTa alloy	normal solidification	mechanical strength	GENERAL ELECTRIC	(105)
TaC fibres	CoCrTa alloy	normal solidification	mechanical strength	GENERAL ELECTRIC	(106)
Interlamellar $\alpha Mg/\beta Li$		normal solidification	—	PRUD'HOMME	(107)
Interlamellar Cd/Zn		pulling a silica slide from the melt (film growth)	—	ALBERS	(108)
Interlamellar α-Al base phase/$CuAl_2$ α-Al base phase/$CuMgAl_2$		—	mechanical strength after aging	RHODES	(109)
Interlamellar $Ag_3Mg/AgMg$		normal solidification	mechanical strength after aging	KIM	(110)
Interlamellar NiAl/V		normal solidification	—	PELLEGRINI	(111)
Interlamellar In_2Bi/In		normal solidification	—	FAVIER	(112)
Interlamellar Zn/Al		normal solidification	—	SINGH	(113)
Interlamellar Ag/Cu Cd/Zn Al/AlAg		normal solidification	—	CANTOR	(114)
Interlamellar Pb/Sn		normal solidification	—	LABULLE	(115)
Interlamellar Pb/Sn		normal solidification	—	FDO	(116)
Interlamellar Al/Al_2Cu		normal solidification	—	RIQUET	(117)
Interlamellar $Co_2Si/CoSiX$ alloy X = W,Mo or both		normal solidification	mechanical strength	LIVINGSTON	(118)
Interlamellar Ni_3Al/Ni_3Cb		normal solidification	oxidation resistance	TARSHIS	(119)
Interlamellar $CoFe_2O_4/BaTiO_3$		normal solidification	piezoelectric, magnetic	PHILIPS	(120)
Interlamellar WXC/W alloy X = IVa metal		normal solidification	wear resistant	AEROJET GENERAL	(121)
Interlamellar $SnSe_2/SnSe$		normal solidification	semiconductor	PHILIPS	(122)
Interlamellar Ni alloy/Cr alloy		normal solidification	mechanical strength	DONER	(123)
Interlamellar $CuAl_2/Al$		normal solidification	mechanical strength	DEAN	(124)

TABLE 12 (End)

MICROSTRUCTURE	MATRIX	SOLIDIFICATION TECHNIQUE	PROPERTIES OR APPLICATION FIELD	AUTHOR(S)	REF.
MgO fibres	MgF_2	normal solidification	optical fibres	PARSONS	(125)
NaF fibres LiF fibres	NaCl NaCl	–	optical fibres	YUE	(126)
Sb lamellae	InSb	zone melting	optical infrared polarizer	CLAWSON	(127)
various halides, nitrates, complex oxides, etc. aligned crystallites	various halides, nitrates, complex oxides, etc.	normal solidification	optical and electromagnetic	LASKO	(128)
PbS fibres	NaCl	normal solidification	optical	PHILIPS	(129)
LiF lamellae	CaF_2	EFG	–	TYCO	(130)
NbC fibres	(Fe,Co,Ni) solid solution	Bridgman technique	Magnetic	BATT *et al.*	(131)
Co fibres	Sm_2Co_{17}	normal solidification	magnetic	SAHM *et al.*	(132)
Co_3Nb lamallae	lamellar Co	normal solidification	magnetic	ARNSON	(133)
Ag platelets	Bi	normal solidification	galvano-thermo-magnetic	DIGGES	(134)
MnBi whiskers	Bi	solidification in magnetic field	magnetic	SAVITSKY	(135)
CeAl alloy fibres	Al	normal solidification	nuclear reactor material	EURATOM	(136)
Interlamellar Pb/Na		normal solidification	superconductor	GUPTA	(137)
MgO rods	ZrO_2	normal solidification	–	KENNARD	(138)
MnO rods	Mn_2SiO_4	EFG	–	FINCH	(139)
Interlamellar NiO/Y_2O_3		floating zone melting	–	BARAILLER	(140)
Fibres of Ta_2Cr; Cr;Ni; or mixtures of Ni and Cr	Al_2O_3,Cr_2O_3; Fe_2O_3,Fe_3O_4; $MgO.Cr_2O_3$; $MgO.Al_2O_3$;CoO. Al_2O_3; mixtures of Al_2O_3, Cr_2O_3, $FeCr_2O_4$ and $FeAl_2O_4$	floating zone melting	oxidation resistant (turbine airfoil)	HULSE	(141)
W fibres Re fibres Mo fibres	UO_2,ThO_2,ZrO_2, MgO,or Cr_2O_3; Cr_2O_3 Cr_2O_3	zone melting	metal conductor in insulator	CLARK	(142)
ZrO_2 lamellae	$CaO-ZrO_2$ phase	floating zone melting	high temperature strength	HULSE	(143)

3.3. Unidirectional Decomposition of Eutectoid Materials.   Eutectoid de-mixing or decomposition is a solid-state reaction analogous to eutectic solidification. By moving an eutectoid composition slowly down a temperature gradient an aligned two-phase structure can be produced.
Aligned eutectoid decomposition avoids crucible contamination, while eutectoid spacings are considerably finer then eutectic spacings produced at the same growth rate because phase separation occurs by solid-state.

VROLIJK (147) described the directional eutectoid decomposition of a Cu-In eutectoid in a Bridgman apparatus, resulting in an interlamellar structure of Cu and Cu_7In_3 crystals.

Duplex crystals of FeAl and $FeAl_2$, produced by directional eutectoid decompositions of a AlFe alloy, have been reported by BASTIN (148).

BUSCHOW (149) obtained closely-spaced parallel lamellae of R_2Co_{17} and R_2Co_7 by decomposition of rare earth-cobalt (RCo_5) compounds, having unusual permanent magnetic properties.

Interlamellar composites of $NiIn/Ni_2In_3$, Cu+29%In/Cu+11%In, Cu+19,6%Al/Cu+30,3%Al, are described by PHILIPS (150). Interlamellar structures are also studied in the CoSi- and Pb-Sn system. Vitreous systems (151) have also been reported, such as $Ca(NO_3)_2$ and $RbNO_3$, $Ca(NO_3)_2$ and $RbNO_3$, $ZnCl_2$ and KCl, Bi_2O_3 and B_2O_3, interlamellar structures being formed on eutectoid decomposition.

LIVINGSTON (152) described the eutectoid decomposition of a unidirectionally solidified CoSi base alloy, obtaining alternating lamellae of a Co_2Si phase and a Co rich phase. He studied also the Cu-11,8%Al system (153) and the Au-40%Ni system (154).

*   *

*

REFERENCES TO CHAPTER 3.

(1)	US 3314825	(NAT. RESEARCH DEVELOPMENT)
(2)	US 3406446	(S. MULDOVAN)
(3)	US 3419952	(GENERAL ELECTRIC)
(4)	US 3936277	(MCDONNEL-DOUGLAS CORP.)
(5)	FR 2083615	(HITACHI)
(6)	US 3691623	(TRW)
(7)	US 3828699	(UK ATOMIC ENERGY AUTHORITY)
(8)	US 3864807	(GRAU)
(9)	FR 1547594	(GENERAL ELECTRIC)
(10)	US 4012824	(FELTEN & GUILLEAUME KABELWERKE)
(11)	FR 2184945	(SIEMENS)
(12)	US 3441392	(MELPAR)
(13)	DE 2745781	(SILAG)
(14)	US 3218697	(HORIZONS INC.)
(15)	GB 733061	(BRITISH THOMSON HOUSTON)
(16)	DE 2250832	(DYNAMIT NOBEL)
(17)	DE 2436951	(R. PRUMMER)
(18)	DE 2928955	(GLYCO-METALLWERKE DAELEN)
(19)	US 3432295	(HITTMAN ASSOC.)
(20)	GB 1172855	(UK ATOMIC ENERGY AUTHORITY)
(21)	US 3994722	(GENERAL DYNAMICS)
(22)	US 3653882	(NASA)
(23)	FR 1328260	(JURID-WERKE)
(24)	JP 8015952	(SUMITOMO)
(25)	S. SKROVANEK *e.a.*, "Am. Ceramic Soc. Bulletin", vol 59 (July 1980) n° 7 p 742 - 745	
(26)	US 4276331	(REPWELL ASSOC.)
(27)	US 4196230	(J.O. GIBSON)
(28)	US 4252588	(SCIENCE APPLICATION)
(29)	US 3553820	(UNION CARBIDE)
(30)	EP 45510	(TOYOTA JIDOSHA K.K.)
(31)	FR 1535660	(THOMSON-HOUSTON)
(32)	US 3668748	(AMERICAN STANDARD)
(33)	FR 2038858	(SOC. IND. DE COMBUSTIBLE NUCLEAIRE)
(34)	FR 1559942	(GENERAL TECHNOLOGY)
(35)	US 3776297	(BATTELLE)
(36)	CH 532969	(BROWN-BOVERI)
(37)	US 3792985	(OWENS CORNING)
(38)	US 3038248	(H. KREMER)
(39)	US 3615277	(UNITED AIRCRAFT)
(40)	FR 2141136	(UNION CARBIDE)
(41)	FR 2191978	(UNITED AIRCRAFT)
(42)	US 3943213	(GREAT LAKES CARBON CORP.)
(43)	US 3505177	(XEROX)

(44) US 3716461   (IQBAL AHMAD)
(45) GB 1155960   (BRITISH IRON & STEEL RESEARCH ASSOCIATION)
(46) DE 1940063   (THE ENGLISH ELECTRIC CY)
(47) US 3498890   (MELPAR)
(48) JP 7968727   (NIPPON CARBON CY)
(49) GB 1224166   (BRISTOL AEROJET)
(50) US 3315663   (J.P. GLASS)
(51) US 3615275   (TEXAS INSTRUMENTS)
(52) US 3794551   (USA SECRETARY OF THE ARMY)
(53) FR 2189207   (DUCOMMUN)
(54) US 4029844   (ATLANTIC RESEARCH)
(55) M. PRUD'HOMME *et al.*, C.R. Acad. Sc. Paris, *272*, C, 10, 902 - 905,
      1971
(56) Y-S SHEN *et al.*, Met. Trans., *1*, 2305 - 2313, 1970
(57) G. HAOUR *et al.*, J. Crystal Growth, *23*, 4, 304 - 106, 1974
(58) H. SPRENGER *et al.*, J. Less-Commen Metals, *34*, 1, 73 - 89, 1974
(59) US 4288259   (D.D. PEARSON *et al.*)
(60) US 4055447   (M.R. JACKSON)
(61) US 4054469   (M.R. JACKSON)
(62) US 4045255   (M.R. JACKSON)
(63) US 4012241   (F.D. LEMKEY)
(64) US 3972748   (J.D. LIVINGSTON)
(65) US 3972746   (J.D. LIVINGSTON)
(66) US 3635769   (B.J. SHAW)
(67) FR 2374426   (UNITED TECHNOLOGIES)
(68) FR 2338332   (UNITED TECHNOLOGIES)
(69) DE 2521563   (SPRENGER)
(70) US 4193822   (COMALCO ALUMINIUM)
(71) J.C. HUBERT *et al.*, J. Crystal Growth, *18*, 3, 241 - 249, 1973
(72) J.C. HUBERT *et al.*, J. Crystal Growth, *18*, 3, 241 - 249, 1973
(73) J.L. WALTER *et al.*, Met. Trans., *4*, 1, 33 - 38, 1973
(74) J.D. LIVINGSTON, Met. Trans., *3*, 12, 3173 - 3176, 1972
(75) L.M. LEVINSON, Appl. Phys. Lett., *21*, 6, 289 - 291, 1972
(76) J.V. GRABEL *et al.*, Met. Trans., *3*, 7, 1973 - 1978, 1972
(77) J.L. WALTER *et al.*, Met. Trans., *1*, 1221 - 1229, 1970
(78) I. NASHIDA *et al.*, J. Crystal Growth, *42*, 540 - 541, 1977
(79) US 3782928   (J.L. WALTER *et al.*)
(80) FR 2109184   (ETAT FRANCAIS)
(81) FR 2159338   (TYCO)
(82) DE 2947917   (US DEPARTMENT OF ENERGY)
(83) FR 2072682   (UNITED AIRCRAFT)
(84) FR 2033230, FR 2011507   (UNITED AIRCRAFT)
(85) J.L. WALTER *et al.*, Met. Trans., *4*, 8, 1775 - 1784, 1973
(86) US 3871835   H. BIBRING *et al.*, Mém. Scient. Rev. Métallurg., *69*, 5,
      341 - 358, 1972
(87) G.D.T. SPILLER *et al.*, J. Crystal Growth, *50*, 445 - 452, 1980
(88) F.M. DUNLEVEY *et al.*, Met. Trans., *5*, 6, 1351 - 1356, 1974
(89) E.R. BUCHANAN *et al.*, Met. Trans., *5*, 6, 1413 - 1422, 1974
(90) E.R. BUCHANAN *et al.*, Met. Trans., *4*, 8, 1895 - 1904, 1973
(91) US 4119458   (W.F. MOORE)
(92) US 4058415   (J.L. WALTER)
(93) US 3785805   (J. VAN DEN BOOMGAARD)
(94) US 3779746, US 3779745   (E. RUDY)
(95) US 3542541   (F.D. LEMKEY)
(96) GB 2047741   (M.F.X. GIGLIOTTI)
(97) GB 2048305   (M.F. HENRY)
(98) GB 1484007   (G.A. CHADWICK)

(99)   GB 1192736   (NATIONAL RESEARCH DEVELOPMENT)
(100) FR 2284684   (ONERA)
(101) FR 2260626   (BROWN BOVERI)
(102) FR 2239537   (ONERA)
(103) FR 2231767   (GENERAL ELECTRIC)
(104) FR 2173094   (PHILIPS)
(105) FR 2136764   (GENERAL ELECTRIC)
(106) FR 2136394   (GENERAL ELECTRIC)
(107) PRUD'HOMME et al., J. Crystal Growth, 19, 1, 65 - 73, 1973
(108) W. ALBERS et al., J. Crystal Growth, 18, 2, 147 - 150, 1973
(109) C.G. RHODES et al., Met. Trans., 3, 7, 1861 - 1868, 1972
(110) Y.G. KIM et al., Met. Trans., 3, 6, 1391 - 1393, 1972
(111) A.W. PELLIGRINI et al., J. Crystal Growth, 42, 536 - 539, 1977
(112) J.J. FAVIER et al., J. Crystal Growth, 38, 109 - 117, 1977
(113) B. SINGH et al., J. Crystal Growth, 37, 301 - 308, 1977
(114) B. CANTOR et al., J. Crystal Growth, 36, 2, 232 - 238, 1976
(115) B. LABULLE et al., J. Crystal Growth, 28, 3, 279 - 287, 1975
(116) E.H FOO et al., J. Crystal Growth, 29, 2, 219 - 222, 1975
(117) J.P. RIQUET et al., J. Crystal Growth, 29, 2, 217 - 218, 1975
(118) US 3972747   (J.D. LIVINGSTON)
(119) US 3783033   (L.A. TARSHIS)
(120) FR 2157401   (PHILIPS)
(121) FR 2033394   (AEROJET GENERAL)
(122) FR 2017385   (PHILIPS)
(123) M. DONER et al., Met. Trans., 5, 2, 433 - 439, 1974
(124) H. DEAN et al., J. Crystal Growth, 21, 51 - 57, 1974
(125) US 4252408, J.D. PARSONS et al., J. Crystal Growth, 55, 470 - 476, 1981
(126) A.S. YUE et al., J. Crystal Growth, 54, 243 - 247, 1981
(127) US 3671102   (A.R. CLAWSON et al.)
(128) US 3505218   (W.R. LASKO et al.)
(129) FR 2157101   (PHILIPS)
(130) FR 2159338   (TYCO)
(131) J.A. BATT et al., J. Appl. Phys., 43, 3, 1295 - 1297, 1972
(132) P.R. SAHM et al., Z. ang. Phys., 30, 1, 95 - 99, 1970
(133) H.L. ARNSON et al., J. Appl. Phys., 45, 8, 3614 - 3616, 1974
(134) T.G. DIGGES et al., Met. Trans., 4, 4, 1169 - 1171, 1973
(135) E.M. SAVITSKY et al., J. Crystal Growth, 52, 519 - 523, 1981
(136) FR 1588139   (EURATOM)
(137) A.D. GUPTA et al., Z. Metallkunde, 66, 5, 265 - 267, 1975
(138) F.L. KENNARD, J. Am. Cer. Soc., 57, 10, 428
(139) C.B. FINCH et al., J. Crystal Growth, 37, 3, 245 - 252, 1977
(140) V. BARAILLER et al., J. Crystal Growth, 51, 3, 632 - 634, 1981
(141) US 4103063   (UNITED TECHNOLOGIES)
(142) US 3796673   (G.W. CLARK et al.)
(143) US 3887384   (UNITED AIRCRAFT CORP)
(144) US 3337295   (J. WEETON)
(145) US 3681063   (NATIONAL RESEARCH AND DEVELOPMENT)
(146) NL 7408602   (US BORAX)
(147) J.W. VROLIJK et al., J. Crystal Growth, 48, 1, 85 - 92, 1980
(148) G.F. BASTIN et al., J. Crystal Growth, 43, 6, 745 - 751, 1978
(149) K.H. BUSCHOW, J. Less-Common Met., 37, 91 - 101, 1974
(150) FR 2101317   (PHILIPS)
(151) FR 2204456   (PHILIPS)
(152) US 3933481   (J.D. LIVINGSTON)
(153) US 3847679   (J.D. LIVINGSTON)
(154) US 3844845   (J.D. LIVINGSTON)

LIST OF
CITED PATENT
DOCUMENTS

LIST OF CITED PATENT DOCUMENTS

FR	1274807	37
	1328260	124
	1423604	5
	1493696	99
	1505474	49
	1511672	49
	1535660	125
	1547594	124
	1551091	50
	1559942	125
	1564841	49
	1568920	49
	1578319	37
	1579111	97
	1583684	49
	1588139	85, 131
	1594182	105
	1598321	53
	1598323	50
	1600655	21
	1603812	19
	2002846	37, 50
	2011507	129
	2011863	113, 114
	2014130	110
	2015951	114
	2017385	130
	2017523	19
	2022113	22
	2022221	17
	2026821	39
	2027513	110
	2029371	31
	2033230	129
	2033394	130
	2036618	50, 53
	2038858	125
	2039169	41
	2039709	21
	2053221	113
	2053222	113
	2056271	17
	2057466	87
	2062169	22
	2064410	51
	2065763	105
	2071294	91
	2072682	129
	2073796	15
	2075256	91
	2075819	64
	2076203	15
	2078869	69
	2080076	87
	2080633	69
	2083615	124
	2084320	92

FR	2084451		87
	2084597		17
	2084811		21
	2086156		105
	2087202		105
	2087413		17
	2087892		64
	2087946		17
	2088130		44
	2091204		105
	2091412		64
	2097792		64
	2098508		22
	2101317		132
	2103411		41
	2107609		15
	2109184		129
	2111009		9
	2111243		89
	2111672		103
	2118974		19
	2123366		21
	2130603		91
	2131858		31
	2133771		51
	2135128		19
	2136394		91, 130
	2136764		130
	2141136	JP 74. 10830	125
	2143124		105
	2144329		104
	2144330		103
	2144760		41
	2150396		64
	2155522		87
	2157101		131
	2157401		130
	2159338	JP 73. 57830	129, 131
	2159660	JP 73. 61728	17
	2165012		89
	2170952		50
	2171414		104
	2173094	JP 73. 97727	130
	2174951	JP 73. 90988	19
	2175882	JP 74. 526	17
	2176041	JP 74. 6007, JP 80.109269	41
	2179760	JP 74. 30626	66
	2182864	JP 73.101407	39
	2184945		124
	2186972	JP 74. 42770	21
	2187417		104
	2189207	JP 80. 51769	125
	2190728	JP 75. 51995	104
	2190764	JP 74. 20206	53
	2190933		87
	2191978		125
	2191996		91

FR	2376831	JP 78. 88813	103
	2377970		66
	2378888	JP 78. 4011	104
	2381000	JP 78.103981	41
	2386890		104
	2391956		104
	2393087	JP 78.147821	14
	2393635	JP 79. 2937	9
	2394507	JP 79. 4895	104
	2396793	JP 79.160427	19
	2398705		104
	2401888		113
	2402631	JP 79. 41913	104
	2404524	JP 79. 61090	104
	2411256		19
	2414574	JP 79.101985	103
	2416270		89
	2416391	JP 79.111060	103
	2421056	JP 79.133581	103
	2424552	JP 79.143242	70
	2424888	JP 79.146824	103
	2427197		103
	2427198	JP 79.159489	103
	2427315		103
	2433003	JP 80.117628	103
	2434964		103
	2441665	JP 80. 69232	91
	2444012		103
	2444650		114
	2446175	JP 80. 93443	103
	2446334	JP 80.128018	43
	2450795	JP 80.142718	39
	2453886	JP 80.144087	19
	2475534	JP 81.169186	110
	2481263		39
	2488244	JP 82. 34085	103
GB	733061		124
	918394		97
	919181		97
	953651		97
	954285		97
	998089		49
	1001003		35
	1030232		37
	1051883		29, 32
	1064271		37
	1069472		35
	1078742		35
	1108633		97
	1108659		29
	1136732		7
	1136922		49
	1141551		49
	1141840		49
	1144033		37

GB	1151464		97
	1155292		37
	1155960		125
	1159210		50
	1163979		99
	1172855		124
	1173740		37
	1177739		19
	1177782		50
	1177854		31
	1190038		65
	1190283		64
	1192736		129
	1198906		99
	1203342		64
	1203343		64
	1204622		50
	1213156		70
	1213867		70
	1215800		31
	1224166		125
	1230970		107
	1236015		107
	1236282		19
	1264476		113
	1301101		23
	1307133		51
	1312258		107
	1320908		22
	1330970		107
	1340069		17
	1343773		107
	1352141		107
	1353384		114
	1354884		41
	1360197		41
	1360198		41
	1360199		41
	1360200		41
	1379547		21
	1400562		70
	1401371	JP 73. 84110	114
	1403862		91
	1403863		110
	1410090		107
	1421672		107
	1434824		107
	1448918		111
	1455331		107
	1457757		107
	1469754		107
	1470292		41
	1475306		107
	1484007		129
	1499457	JP 76. 70323	17
	1500675	JP 77. 31124	14
	1505095		14

JP	81.110733	53
	81.129668	114
NL	19624	29
	6604168	7
	7100743	31
	7112396	7
	7408602	126
	7507558	31
SU	356264	114
	374256	114
	374258	114
	380615	111
	381645	111
	381647	111
	381650	111
	386874	111
	390049	111
	392046	111
	393251	111
	395342	111
	414232	111
	415247	111
	422705	111
	458535	111
	477974	111
	478818	111
	483378	111
	492506	99
	528286	111
	553228	111
	833850	111
	833872	99
US	1879336	7
	2886866	7
	2976590	7
	3023029	22
	3038248	125
	3082051	37
	3082054	37
	3082055	37
	3082099	37
	3082103	37
	3096144	37
	3180741	37
	3199331	7
	3214805	5
	3216076	7
	3218697	124
	3233985	97
	3264388	7
	3269802	49

US	3271173	37
	3294880	49
	3311481	37
	3311689	37
	3314825	124
	3315663	125
	3321285	97
	3322865	37
	3337295	126
	3362803	5
	3370923	50
	3384578	97
	3385915	37
	3386840	97
	3386918	97
	3394213	5
	3398013	49
	3406446	124
	3416944	99
	3416951	49
	3416953	37
	3419952	124
	3427185	97
	3428719	37, 50
	3429722	50
	3432295	124
	3433725	49
	3441392	124
	3451840	29
	3462289	99
	3466352	7
	3488151	19
	3492119	92
	3493431	69
	3496078	71
	3498890	125
	3501491	22
	3503765	37
	3505014	64
	3505127	69
	3505177	125
	3505218	131
	3510275	85
	3519492	63
	3525589	65
	3527564	14
	3536519	62
	3542541	129
	3549413	50
	3550213	110
	3553820	125
	3565749	37, 50
	3572286	31
	3573961	22
	3575789	110
	3582271	64
	3592595	15

US	3801351		21
	3806489		22
	3808015		39
	3811927		31, 50
	3811930		31
	3813219		15
	3814377		15
	3814782		43
	3816598		21
	3821013		22
	3828699		111, 124
	3832297		21
	3833389		111
	3833402		22
	3837904		23
	3838488		7
	3840647		63
	3840649		19
	3841079		14
	3843762		9
	3844822		23
	3844845		132
	3846527		41
	3846833		14
	3847558		85
	3847679		132
	3849181	JP 74. 35627	44
	3850689		51
	3853600		23
	3853610		22
	3853688		39
	3854518	JP 75. 97524	9
	3856593		109
	3859043		14
	3860443		91
	3860529		41
	3861947		44
	3862658		9
	3864807		124
	3865917		43
	3867491		105
	3868230		50
	3870444		21
	3871439	JP 75. 46546	9
	3875296		67
	3887384	JP 75.115209	114, 131
	3887722		31
	3888661		89
	3897582		109
	3899574		19
	3900556		17
	3900626		99
	3900675		109
	3901719		111
	3903220	JP 74. 86636	17
	3903248		19
	3903323		31

US	3903347		31
	3908061		22
	3909278	JP 75. 97598	43
	3914395		109
	3914500		99
	3915663		62
	3917776		17
	3917783		31
	3917884		109
	3922411		101
	3925133		109
	3927157		109
	3927180		66
	3931392		21
	3933481		132
	3935354		109
	3936277		124
	3936522		109
	3938964		85
	3943213		105, 125
	3944640		70
	3947562		64
	3948669		111
	3949126		109
	3953174	JP 76.132184	70
	3953561		39
	3953647		89
	3957716	JP 75. 59589	22
	3959453	JP 75. 85597	66
	3960592		22
	3961105	JP 74. 30626	66
	3966887	JP 73. 61728	17
	3969545	JP 75. 25243	62
	3971840	JP 74.125631	53
	3972746		127
	3972747		130
	3972748		127
	3972984	JP 76. 70324	17
	3974264	JP 75. 89635, JP 78. 86717, JP 78. 86718, JP 79. 11330	17
	3976746		14
	3982955	JP 73. 75823	41
	3989802		21
	3991248	JP 80. 51769	104, 113
	3992160		92
	3992498		41
	3994722		124
	3995024		19
	4002426		15
	4002725		66
	4004053		15
	4005172		22
	4008299		43
	4009248	JP 76.116225	15
	4010233		41
	4012204	JP 76. 70116	92
	4012241		127

US			
4012824			124
4016247			19
4020145			17
4024227	JP 76. 55425		15
4026745			109
4029844		109,	125
4031288	JP 73. 56288		15
4031851			31
4036599			92
4039341			17
4040890			70
4041116			109
4045255			127
4045597			31
4047965	JP 77.137030		43
4048953	JP 76. 7295		103
4049338			62
4054469	JP 77.155125		127
4055447			127
4056874			89
4058415			129
4058699			70
4064207			109
4067742			101
4071594			39
4075276	JP 75.120499	54,	113
4082864			89
4094690	JP 74.132200		41
4097294			54
4100004			17
4100044			71
4101615	JP 74.108325, JP 75. 12335		43
4103063		99,	131
4104045	JP 77. 10312		41
4104354	JP 74. 99632		109
4104355			39
4104395		41,	59
4104445			51
4107352		50,	53
4115527			19
4116689			91
4117565	JP 78. 29214		91
4119458			129
4123832			109
4125406	JP 75. 39311		43
4131708			107
4141726			89
4142008			31
4147538	JP 78. 83884		91
4148671			85
4150708			9
4152149	JP 75.109903		43
4152381			109
4152408	JP 79. 82397		59
4152482			109
4155781	JP 78. 31987		69
4157409			89

US	4157729		9
	4162301		51, 53
	4163583		31
	4164601		109
	4165355		99, 113
	4166145		109
	4166147	JP 75. 10307	43
	4175153	JP 79.151622	41, 51
	4180399	JP 78. 40612	91
	4180409	JP 77. 10312	41
	4180428		113
	4191561		71
	4193822		127
	4193828		107
	4196230		113, 124
	4201611		107
	4209008		71
	4209560		99, 113
	4216262		22
	4222977	JP 79.151619	39, 51
	4223075		91
	4238433		114
	4240835	JP 81. 92181	114
	4250131		43
	4252408		131
	4252588		107, 124
	4253731		71
	4256792		113
	4259125		9
	4260007		9
	4263367		110, 111
	4265872		39, 67
	4265968		110, 111
	4268562		110, 111
	4275095		113
	4276331		124
	4277325	JP 80.144087, JP 82. 2393	19
	4278729		113
	4284610		113
	4284664		114
	4288259		127
	4294788	JP 81. 92181	114
	4298558	JP 82. 34129	53
	4298559	JP 82. 34132	53
	4301136	JP 76.116424	17
	4303631	JP 82. 42924	19
	4310651		53
	4331739		9
	4340619	JP 82.117532	54
	4341826	JP 81.169186	110, 111
WO	81/ 523		41
	81/2733		54
	81/2734		54
	81/2755		54

LIST OF
PATENTEES

List of Patentees

AEROJET GENERAL	FR	2033394
AEROSPACE CORPORATION	US	4223075
AGENCY OF INDUSTRIAL SCIENCE & TECHNOLOGY	DE	2808373
AGENCY OF INDUSTRIAL SCIENCE & TECHNOLOGY	JP	73. 48720
AGENCY OF INDUSTRIAL SCIENCE & TECHNOLOGY	JP	77.114727
AHMAD, I.	US	3607451
AHMAD, I.	US	3716461
AKZO	FR	2446334
ALLIED CHEMICAL CORPORATION	FR	2281434
ALLIED CHEMICAL CORPORATION	US	3862658
ALLIED CHEMICAL CORPORATION	US	4260007
ALLIED CHEMICAL CORPORATION	US	4331739
ALUMINIUM COMPANY OF AMERICA	US	3705223
ALUMINIUM COMPANY OF AMERICA	US	4071594
AMERICAN STANDARD	US	3668748
ANDERSON, R.H.	US	4253731
ANVAR	FR	2259916
ARMINES	FR	2080076
ARMINES	FR	2170952
ARTHUR, J.R.	US	3635753
ARTHUR D. LITTLE	US	3944640
ARTHUR D. LITTLE	US	4058699
ASAHI CHEMICAL INDUSTRY	JP	79. 84000
ASAHI GLASS	JP	75. 27479
ASANO, T.	JP	75.136306
ASSOCIATED ELECTRICAL INDUSTRIES	GB	1108633
ATLANTIC RESEARCH	US	3717419
ATLANTIC RESEARCH	US	3897582
ATLANTIC RESEARCH	US	3900675
ATLANTIC RESEARCH	US	3925133
ATLANTIC RESEARCH	US	3935354
ATLANTIC RESEARCH	US	4029844
AUDI NSU AUTO UNION	DE	2912786
AVCO CORPORATION	FR	2337214
AVCO CORPORATION	GB	2080781
AVCO CORPORATION	GB	2081695
AVCO CORPORATION	NL	7100743
AVCO CORPORATION	US	3868230
AVCO CORPORATION	US	3914395
AVCO CORPORATION	US	3922411
AVCO CORPORATION	US	4045597
AVCO CORPORATION	US	4142008
AVCO CORPORATION	US	4163583
AVCO CORPORATION	US	4165355
AVCO CORPORATION	US	4209560

THE BABCOCK & WILCOX CORPORATION	FR	2272968
THE BABCOCK & WILCOX CORPORATION	US	3322865
THE BABCOCK & WILCOX CORPORATION	US	3503765
BATTELLE DEVELOPMENT CORPORATION	US	3776297
BATTELLE DEVELOPMENT CORPORATION	US	3871439
BAYER	DE	2236078
BAYER	DE	2702097
BAYER	DE	2702100
BAYER	DE	2752367
BAYER	FR	2039169
BAYER	FR	2064410
BAYER	FR	2078869
BAYER	FR	2080633
BAYER	FR	2084597
BAYER	FR	2103411
BAYER	FR	2133771
BAYER	FR	2175882
BAYER	FR	2190764
BAYER	FR	2197829
BAYER	FR	2203777
BAYER	FR	2229788
BAYER	FR	2377970
BAYER	US	3846527
BAYER	US	3982955
BAYER	US	4010233
BAYER	US	4104045
BAYER	US	4180409
THE BENDIX CORPORATION	FR	2143124
BERISFORD, R.	US	3728443
BICKERDIKE, R.L.	DE	1571320
BITZER, D.	DE	2430719
BJORKSTEN RESEARCH LABORATORIES	JP	80. 23007
BJORKSTEN RESEARCH LABORATORIES	US	4104355
BRIDENBAUGH,P.M.	US	4002725
BRISTOL AEROJET	GB	1224166
BRITISH IRON & STEEL RESEARCH ASSOCIATES	GB	1155960
BRITISH NUCLEAR FUELS	FR	2187417
BROWN BOVERI	CH	532969
BROWN BOVERI	DE	1928373
BROWN BOVERI	FR	2075256
BROWN BOVERI	FR	2260626
BRUBAKER, B.D.	US	3711599
BRUBAKER, B.D.	US	3875296
BRUNSWICK CORPORATION	GB	1136732
BRUNSWICK CORPORATION	GB	1512811
BRUSH WELLMAN	FR	2196393
BURNETT, P.	US	3630691
BURRUS, C.A.	US	4040890
CABOT CORPORATION	US	3715791
CAMAHORT, J.	US	4031851
LE CARBONE-LORRAINE	FR	1493696
LE CARBONE-LORRAINE	FR	2065763
LE CARBONE-LORRAINE	FR	2398705
LE CARBONE-LORRAINE	GB	2099365
THE CARBORUNDUM CORPORATION	EP	7485

THE CARBORUNDUM CORPORATION	FR	1594182
THE CARBORUNDUM CORPORATION	FR	2182864
THE CARBORUNDUM CORPORATION	FR	2359085
THE CARBORUNDUM CORPORATION	FR	2404524
THE CARBORUNDUM CORPORATION	GB	1307133
THE CARBORUNDUM CORPORATION	US	3429722
THE CARBORUNDUM CORPORATION	US	3462289
THE CARBORUNDUM CORPORATION	US	3620780
THE CARBORUNDUM CORPORATION	US	3630766
THE CARBORUNDUM CORPORATION	US	3668059
THE CARBORUNDUM CORPORATION	US	3689220
THE CARBORUNDUM CORPORATION	US	3712428
THE CARBODUNDUM CORPORATION	US	3759353
THE CARBORUNDUM CORPORATION	US	3772115
THE CARBORUNDUM CORPORATION	US	3867491
THE CARBORUNDUM CORPORATION	US	3903220
THE CARBORUNDUM CORPORATION	US	3971840
THE CARBORUNDUM CORPORATION	US	4075276
CELANESE CORPORATION	EP	15729
CELANESE CORPORATION	FR	2002846
CELANESE CORPORATION	FR	2076203
CELANESE CORPORATION	US	3592595
CELANESE CORPORATION	US	3647770
CELANESE CORPORATION	US	3650668
CELANESE CORPORATION	US	3656882
CELANESE CORPORATION	US	3656883
CELANESE CORPORATION	US	3656903
CELANESE CORPORATION	US	3657409
CELANESE CORPORATION	US	3677705
CELANESE CORPORATION	US	3708326
CELANESE CORPORATION	US	3723605
CELANESE CORPORATION	US	3723607
CELANESE CORPORATION	US	3729549
CELANESE CORPORATION	US	3745104
CELANESE CORPORATION	US	3754957
CELANESE CORPORATION	US	3762941
CELANESE CORPORATION	US	3767774
CELANESE CORPORATION	US	3813219
CELANESE CORPORATION	US	3821013
CELANESE CORPORATION	US	3841079
CELANESE CORPORATION	US	3844822
CELANESE CORPORATION	US	3846833
CELANESE CORPORATION	US	3853600
CELANESE CORPORATION	US	3900556
CELANESE CORPORATION	US	3903248
CELANESE CORPORATION	US	4002426
CELANESE CORPORATION	US	4004053
CELANESE CORPORATION	US	4009248
CELANESE CORPORATION	US	4020145
CELANESE CORPORATION	US	4056874
CERTAIN-TEED CORPORATION	US	4152408
CHADWICK, G.A.	GB	1484007
CHARBONNAGES DE FRANCE	FR	2087413
CHARBONNAGES DE FRANCE	FR	2159660
CHARBONNAGES DE FRANCE	US	3966887
CHEMOTRONICS	DE	2559608
CHENOT, C.F.	US	3927180

CHIAKA ASADA	US	4117565
CHIAKA ASADA	US	4180399
CLAIRAIRE	GB	1505095
CLEVITE CORPORATION	US	3216076
CLIFTON, R.A.	US	3525589
CLARK, G.W.	US	3796673
CLAWSON, A.R.	US	3671102
COAL INDUSTRIES	FR	2111672
COAL INDUSTRIES	FR	2118974
COAL INDUSTRIES	FR	2296592
COAL INDUSTRIES	GB	1434824
COMALCO ALUMINIUM	US	4193822
COMMISSARIAT A L'ENERGIE ATOMIQUE	FR	2087202
COMMISSARIAT A L'ENERGIE ATOMIQUE	FR	2427315
COMMISSARIAT A L'ENERGIE ATOMIQUE	FR	2433003
CONRAD, R.W.	US	3607054
C. CONRADTY NUERNBERG GmbH	FR	2376831
C. CONRADTY NUERNBERG GmbH	FR	2391956
CONSORTIUM FUER ELECTROCHEMISCHE INDUSTRIE	DE	1696101
CORBETT ASSOCIATES	US	3466352
CORNING GLASS WORKS	US	3901719
CORNING GLASS WORKS	US	3948669
COURTAULDS	FR	2039709
COURTAULDS	FR	2087946
COURTAULDS	GB	1236015
COURTAULDS	GB	1379547
DANNOEHL, W.	CH	588416
DANNOEHL, W.	FR	1423604
DANNOEHL, W.	FR	2084320
DANNOEHL, W.	US	3362803
DEITZ, V.	US	3931392
DENKI	JP	77. 32906
DENKI	JP	80. 20239
DENKI	JP	80. 45808
DENKI	JP	80. 45809
DIEPERS, H.	US	4155781
DOW CHEMICAL	US	3840649
DOW CHEMICAL	US	3853610
DOW CHEMICAL	US	3908061
DOW CORNING	FR	1564841
DOW CORNING	FR	1568920
DOW CORNING	GB	1136922
DOW CORNING	GB	1141551
DOW CORNING	GB	1141840
DOW CORNING	GB	2021545
DOW CORNING	GB	2081286
DOW CORNING	GB	2081288
DOW CORNING	GB	2081289
DOW CORNING	US	4298558
DOW CORNING	US	4298559
DOW CORNING	US.	4310651
DOW CORNING	US	4340619
DUCOMMUN	FR	2189207
DUCOMMUN	FR	2289459
DUCOMMUN	GB	1163979

DUCOMMON	US	3991248
DUNLOP	GB	1469754
DUNLOP	GB	1549687
DUNLOP	FR	2062169
DUNLOP	FR	2260726
DUNLOP	FR	2270203
DUNLOP	FR	2414574
DU PONT	FR	2026821
DU PONT	FR	2237841
DU PONT	FR	2381000
DU PONT	US	3214805
DU PONT	US	3808015
DU PONT	US	3849181
DU PONT	US	3853688
DU PONT	US	3953561
DU PONT	US	4012204
DU PONT	US	4036599
DU PONT	WO	81/ 523
DYNAMIT NOBEL	DE	2250832
EATON CORPORATION	EP	37104
EBERL, J.J.	FR	2179760
EBERL, J.J.	US	3961105
ECOLE NATIONALE SUPERIEURE DES MINES DE PARIS ET SNECMA	FR	1583684
ECOLE NATIONALE SUPERIEURE DES MINES DE PARIS ET SNECMA	FR	1598321
ECOLE NATIONALE SUPERIEURE DES MINES DE PARIS ET SNECMA	FR	1598323
ELBAN, W.L.	US	3833402
ELLIS, C.H.	DE	1964991
THE ENGLISH ELECTRICITY COMPANY	DE	1940063
ENGLISH ELECTRIC	GB	1320908
ETAT FRANCAIS	FR	2029371
ETAT FRANCAIS	FR	2109184
EURATOM	FR	1588139
EURATOM	FR	2084451
EURATOM	FR	2190933
EURATOM	FR	2386890
EVANS, C.C.	US	3677713
EXXON RESEARCH	EP	34910
EXXON RESEARCH	FR	2396793
EXXON RESEARCH	FR	2453886
EXXON RESEARCH	US	4277325
LES FABRIQÚES D'ASSORTIMENTS REUNIES	CH	586166
FELDMÜHLE	DE	1471152
FELDMÜHLE	DE	1671891
FELTEN & GUILLAUME KABELWERKE	US	4012824
FIBER MATERIALS	FR	2323527
FIBER MATERIALS	GB	2084978
FIBER MATERIALS	US	3860443
FIBER MATERIALS	US	3917884
FIBER MATERIALS	US	3949126
FIBER MATERIALS	US	4082864
FIBER MATERIALS	US	4131708
FIBER MATERIALS	US	4193828
FITZER, E.	DE	2103908

FITZER, E.	DE	2113129
FLETCHER, J.C.	US	3706583
FLETCHER, J.C.	US	4067742
FMC CORPORATION	GB	1064271
FMC CORPORATION	GB	1155292
FMC CORPORATION	GB	1173740
FMC CORPORATION	US	3428719
FMC CORPORATION	US	3565749
FOLEY, F.W.	US	1879336
FORDATH	EP	27251
FORDATH	GB	1230970
FORDATH	GB	1330970
FUJI	EP	54437
FUJIKI	US	4265872
GANTMAN, S.A.	SU	833850
GENERAL DYNAMICS	US	3994722
GENERAL DYNAMICS	US	4116689
GENERAL ELECTRIC (GE)	DE	2208212
GE	FR	1547594
GE	FR	2136394
GE	FR	2136764
GE	FR	2227244
GE	FR	2231767
GE	FR	2373348
GE	FR	2444650
GE	GB	919181
GE	GB	1190283
GE	GB	1203342
GE	GB	1203343
GE	US	3419952
GE	US	3505014
GE	US	3620836
GE	US	3668006
GE	US	3668062
GE	US	3764662
GE	US	3870444
GE	US	3899574
GE	US	4123832
GE	US	4238433
GE	US	4240835
GE	US	4284664
GE	US	4294788
GENERAL MOTORS	US	3717443
GENERAL TECHNOLOGIES (GT)	FR	1559942
GT	GB	1204622
GT	US	3549413
GT	US	3607367
GT	US	3741797
GIBSON, J.O.	US	4196230
GIBSON, J.O.	US	4278729
GIGLIOTTI, M.F.X.	GB	2047741
GLASS, J.P.	US	3315663
GLASS, J.P.	US	3536519
GLASS, J.P.	US	3615258
GLASS, J.P.	US	3915663

GLYCO-METALLWERKE DAELEN	DE	2928955
GOODRICH	EP	29851
GOODRICH	US	3936552
GOODYEAR AEROSPACE	FR	2313601
GOODYEAR AEROSPACE	FR	2416391
GOODYEAR TIRE & RUBBER	FR	2144329
GOODYEAR TIRE & RUBBER	FR	2144330
GOODYEAR TIRE & RUBBER	FR	2225654
GRAU	US	3864807
GREAT LAKES CARBON (GLC)	FR	2056271
GLC	US	3746560
GLC	US	3776829
GLC	US	3811927
GLC	US	3837904
GLC	US	3943213
GLC	US	3989802
GLC	US	4041116
GLC	US	4216262
GRINDSTAFF, L.I.	US	3787541
GRINSHAW, R.W.	US	3947562
GROUP SERVICES	GB	954285
GROUP SERVICES	GB	998089
GTE SYLVANIA	US	4150708
GTE SYLVANIA	US	4157729
HAVEG INDUSTRIES	US	3758352
HAVEG INDUSTRIES	US	3796616
HAVEG INDUSTRIES	US	3856593
HAWKINS	US	3573961
HENRY, M.F.	GB	2048305
HERCULES	US	3832297
HERCULES	US	3957716
HITACHI	DE	2164568
HITACHI	DE	2649704
HITACHI	FR	2083615
HITACHI	JP	74.121734
HITCO	US	3311481
HITCO	US	3416953
HITCO	US	3728423
HITCO	US	3927157
HITCO	US	3976746
HITCO	US	4026745
HITCO	US	4104354
HITCO	US	4164601
HITCO	US	4166145
HITTMAN ASSOCIATES	US	3432295
HOLLANDER, E.F.	US	3664813
HONEYWELL	US	4256792
HOUGH, R.L.	US	3370923
HOUGH, R.L.	US	3416951
HOUGH, R.L.	US	3433725
HOUGH, R.L.	US	3451840
HORIZONS	US	3082051
HORIZONS	US	3082054
HORIZONS	US	3082055
HORIZONS	US	3082099

HORIZONS	US	3082103
HORIZONS	US	3096144
HORIZONS	US	3180741
HORIZONS	US	3218697
HORIZONS	US	3269802
HORIZONS	US	3271173
HORIZONS	US	3311689
HULSE, C.O.	US	3887384
HUML, J.O.	US	3519492
HUML, J.O.	US	3598526
HUSSEY, C.L.	US	4100044
HYFIL	FR	2196966
IMPERIAL CHEMICAL INDUSTRIES (ICI)	FR	2144760
ICI	FR	2176041
ICI	FR	2213253
ICI	GB	1354884
ICI	GB	1360197
ICI	GB	1360198
ICI	GB	1360199
ICI	GB	1360200
ICI	GB	1470292
ICI	GB	2059933
ICI	US	3960592
ICI	US	3992498
ICI	US	4005172
ICI	US	4008299
ICI	US	4094690
IMPERIAL METAL INDUSTRIES	FR	2155522
INSTITUT FIZIKI	FR	2416270
INSTITUT FIZIKI	GB	1448918
JACKSON, M.R.	US	4045255
JACKSON, M.R.	US	4054469
JACKSON, M.R.	US	4055447
JAPAN EXLAN	DE	2361190
JAPAN EXLAN	DE	2407372
JAPAN EXLAN	DE	2420101
JAPAN EXLAN	DE	2506344
JAPAN EXLAN	FR	2236034
JAPAN EXLAN	GB	1499457
JAPAN EXLAN	GB	1500675
JAPAN EXLAN	GB	2011364
JAPAN EXLAN	US	4024227
JOHNS-MANVILLE	FR	2348899
JOHNS-MANVILLE	FR	2450795
JOSEPH LUCAS	FR	2015951
JOSEPH LUCAS	FR	2053221
JOSEPH LUCAS	FR	2053222
JURID-WERKE	FR	1328260
KAISER ALUMINUM	US	3264388
KANEBO	FR	2394507
KANEBO	FR	2402631

KANEGAFUCHI	JP	77.156198
KANEGAFUCHI	JP	77.156199
KANEGAFUCHI BOSEKI	DE	2158798
KANEGAFUCHI BOSEKI	FR	2075819
KANEGAFUCHI BOSEKI	FR	2087892
KARPINOS	SU	380615
KARPINOS	SU	381645
KARPINOS	SU	381647
KARPINOS	SU	381650
KARPINOS	SU	390049
KARPINOS	SU	392046
KARPINOS	SU	393251
KARPINOS	SU	395342
KARPINOS	SU	414232
KARPINOS	SU	415247
KARPINOS	SU	477974
KARPINOS	SU	478818
KARPINOS	SU	483378
KARPINOS	SU	492506
KARPINOS	SU	553228
KELLSEY, R.H.	US	3658469
KENNECOTT	EP	31656
KENNECOTT	US	4284610
KENNECOTT	WO 81/	2733
KENNECOTT	WO 81/	2734
KENNECOTT	WO 81/	2755
KOPPERS	FR	2135128
KOPPERS	FR	2219906
KREMER, H.	US	3038248
KROCHMAL, J.J.	US	3398013
KULAKOV, V.V.	FR	2434964
KUREHA	DE	1930713
KUREHA	DE	2165029
KUREHA	DE	2205122
KUREHA	FR	2086156
KUREHA	FR	2190728
KUREHA	JA	72. 22679
KUREHA	US	3639953
KUREHA	US	3666417
KUREHA	US	4016247
KUREHA	US	4115527
KUROSAKI REFRACTORIES	JP	81.129668
LABELLE, H.E.	US	3627574
LABELLE, H.E.	US	3650703
LABELLE, H.E.	US	3953174
LASKO, W.R.	US	3505218
LEE, S.A.	US	3709981
LEMKEY, F.D.	US	3542541
LEMKEY, F.D.	US	4012241
LEMKEY, F.D.	US	4209008
LESOVOI, M.V.	SU	458535
LEVITT, A.P.	US	3888661
LEVITT, A.P.	US	4157409
LINDEN, H.	DE	2615523
LIVINGSTON	US	3844845

LIVINGSTON	US	3847679
LIVINGSTON	US	3933481
LIVINGSTON	US	3972746
LIVINGSTON	US	3972747
LIVINGSTON	US	3972748
LOCKHEED	US	3634132
LOCKHEED	US	3816598
LONZA-WERKE	CH	514500
LONZA-WERKE	FR	2091412
LONZA-WERKE	FR	2097792
LONZA-WERKE	FR	2150396
MARVALAUD	FR	1168521
MARVALAUD	US	2886866
MARVALAUD	US	2976590
McDONNEL DOUGLAS	US	3736159
McDONNEL DOUGLAS	US	3766000
McDONNEL DOUGLAS	US	3936277
MELPAR	US	3441392
MELPAR	US	3498890
MICHELIN	FR	2367564
MICHELIN	FR	2393635
MIMURA, Y.	FR	2424552
MINAGAWA, S.	US	3582271
MINESOTA MINING & MANUFACTURING (MMM)	US	3321285
MMM	US	3760049
MMM	US	3793041
MMM	US	3795524
MMM	US	3909278
MMM	US	4031288
MMM	US	4047965
MMM	US	4125406
MMM	US	4166147
MITSUBISHI RAYON	FR	2231414
MITSUBISHI RAYON	JA	74. 71218
MITSUBISHI RAYON	US	3917776
MONSANTO	DE	2013913
MONSANTO	EP	47640
MONSANTO	FR	2017523
MONSANTO	FR	2107609
MONSANTO	FR	2111009
MONSANTO	GB	1236282
MONSANTO	NL	6604168
MONSANTO	US	3386840
MONSANTO	US	3627466
MONSANTO	US	3627570
MONSANTO	US	3720741
MONSANTO	US	3727292
MONSANTO	US	3814377
MONSANTO	US	3854518
MONSANTO	US	4104445
MONSANTO	US	4175153
MONSANTO	US	4222977
MOORE, W.F.	US	4119458
MORGANITE MODMUR	FR	2084811
MORGANITE MODMUR	FR	2378888
MULDOVAN, S.	US	3406446

NAGAKOME, Y.	GB	2016731
NASA	FR	2130603
NASA	US	3653882
NATIONAL BERYLLIA CORPORATION	GB	1001003
NATIONAL RESEARCH	DE	1949128
NATIONAL RESEARCH	DE	2045680
NATIONAL RESEARCH	FR	2011863
NATIONAL RESEARCH	FR	2022221
NATIONAL RESEARCH	FR	2098508
NATIONAL RESEARCH	FR	2204570
NATIONAL RESEARCH	FR	2328723
NATIONAL RESEARCH	FR	2328787
NATIONAL RESEARCH	GB	1192736
NATIONAL RESEARCH	GB	1198906
NATIONAL RESEARCH	GB	1340069
NATIONAL RESEARCH	GB	1352141
NATIONAL RESEARCH	US	3199331
NATIONAL RESEARCH	US	3314825
NATIONAL RESEARCH	US	3681063
NATIONAL RESEARCH	US	3728168
NATIONAL RESEARCH	US	4039341
NICKL, J.	US	3607067
NIPPON ASBESTOS	JP	81. 9427
NIPPON CARBON	DE	2623968
NIPPON CARBON	DE	2659374
NIPPON CARBON	JP	79. 68727
NIPPON CARBON	US	3972984
NIPPON SEIZEN	DE	2339466
NIPPON SEIZEN	FR	2192882
NIPPON SEIZEN	NL	7112396
NIPPON SEIZEN	US	3643304
NITTO BOSEKI	US	3639140
NITTO BOSEKI	US	3661616
NORTH AMERICAN ROCKWELL	FR	1579111
NORTHROP	NL	7507558
NORTHROP	US	3917783
NORTON	FR	1274807

OFFICE NATIONAL D'ETUDES ET DE RECHERCHE AEROSPATIALE (ONERA)	FR	2071294
ONERA	FR	2239537
ONERA	FR	2284684
ONERA	FR	2411256
ONERA	FR	2441665
OWENS CORNING	US	3575789
OWENS CORNING	US	3607608
OWENS CORNING	US	3792985
OWENS CORNING	US	3992160

PARSONS, J.D.	US	4252408
PEARSON, D.D.	US	4288259
PEPPER,R.T.	US	3770488
PFIZER	US	4048953
PHILIPS	FR	2017385
PHILIPS	FR	2101317
PHILIPS	FR	2157101
PHILIPS	FR	2157401

PHILIPS	FR	2173094
PHILIPS	FR	2204456
PHILIPS	FR	2369230
PHILIPS	GB	1213156
PHILIPS	GB	1213867
PHILIPS	GB	1400562
PHILIPS	NL	19624
THE PLESSEY COMPANY	GB	918394
THE PLESSEY COMPANY	GB	953651
THE PLESSEY COMPANY	GB	1264476
THE PLESSEY COMPANY	GB	1312258
THE PLESSEY COMPANY	US	3720575
PRUMMER, R.	DE	2436951
QUINLAN, K.P.	US	4191561
RASHID, M.S.	US	3023029
REPWELL ASSOCIATES	US	4276331
THE RESEARCH INSTITUTE FOR IRON, STEEL AND OTHER METALS OF THE TOHOKU UNIVERSITY	FR	2308590
THE RESEARCH INSTITUTE FOR IRON, STEEL AND OTHER METALS OF THE TOHOKU UNIVERSITY	FR	2308650
THE RESEARCH INSTITUTE FOR IRON, STEEL AND OTHER METALS OF THE TOHOKU UNIVERSITY	FR	2329611
THE RESEARCH INSTITUTE FOR IRON, STEEL AND OTHER METALS OF THE TOHOKU UNIVERSITY	FR	2334757
THE RESEARCH INSTITUTE FOR IRON, STEEL AND OTHER METALS OF THE TOHOKU UNIVERSITY	FR	2345477
THE RESEARCH INSTITUTE FOR IRON, STEEL AND OTHER METALS OF THE TOHOKU UNIVERSITY	FR	2347463
THE RESEARCH INSTITUTE FOR IRON & STEEL	US	4141726
THE RESEARCH INSTITUTE FOR SPECIAL INORGANIC MATERIALS	DE	3018465
RHONE-POULENC	FR	2186972
RHONE-PROGIL	US	3806489
RIBBON TECHNOLOGY	US	4259125
RICE, R.W.	US	4097294
ROEHR PRODUCTS	US	3394213
ROLLS-ROYCE	FR	1603812
ROLLS-ROYCE	FR	2022113
ROLLS-ROYCE	FR	2027513
ROLLS-ROYCE	GB	1069472
ROLLS-ROYCE	GB	1078742
ROLLS-ROYCE	GB	2005237
ROSENBERG, R.	US	3492119
ROSS, J.M.	US	3501491
RUDY, E.	US	3779745
RUDY, E.	US	3779746
SAKUHANA, M.	JP	77. 47052
SAMSONOV, G.V.	SU	356264
SAMSONOV, G.V.	SU	374256
SAMSONOV, G.V.	SU	374258
SAMSONOV, G.V.	SU	386874
SAMSONOV, G.V.	SU	422705
SCHADE, W.	DE	2250116

SCHEICHER, H.	DE	2711219
SCHMIDT, R.	US	3938964
SCHNABEL, R.	DE	440745
SCHUNK & EBE	DE	2714364
SCHWOPE	US	3510275
SCIENCE APPLICATIONS	US	4252588
THE SECRETARY OF STATE FOR DEFENCE	GB	1301101
THE SECRETARY OF STATE FOR DEFENCE	GB	1457757
THE SECRETARY OF STATE FOR DEFENCE	GB	2014972
THE SECRETARY OF THE NAVY	US	3859043
SEGURICUM	GB	2014971
SEGURICUM	US	4100004
SEISHI YAJIMA	DE	2647862
SEISHI YAJIMA	US	4147538
SHAW, B.J.	US	3635769
SHELL	GB	1475306
SHELL	GB	1546802
SHEVCHENKO, A.V.	SU	833872
SIEMENS	FR	2184945
SIGRI ELEKTROGRAPHIT GmbH	DE	2220614
SIGRI ELEKTROGRAPHIT GmbH	FR	2171414
SILAG	DE	2745781
SLEIGH, G.	US	3843762
SLOCUM, R.E.	US	3969545
SLOCUM, R.E.	US	4049338
SOCIETE DE FABRICATION D'ELEMENTS CATALYTIQUES	FR	2088130
SOCIETE EUROPEENNE DE PROPULSION (SEP)	EP	32097
SEP	EP	32858
SEP	EP	57637
SEP	FR	2276913
SEP	FR	2276916
SEP	FR	2334495
SEP	FR	2401888
SEP	FR	2421056
SEP	FR	2424888
SEP	FR	2427197
SEP	FR	2427198
SEP	FR	2444012
SEP	FR	2446175
SOCIETE EUROPEENNE DES PRODUITS REFRACTAIRES	FR	2481263
SOCIETE INDUSTRIELLE DE COMBUSTIBLE NUCLEAIRE	FR	2038858
SNECMA	FR	2131858
SOCIETE NATIONALE DES POUDRES ET EXPLOSIFS (SNPE)	EP	18260
SNPE	FR	2243907
SPACE AGE MATERIALS	FR	1511672
SPACE AGE MATERIALS	US	3294880
SPRENGER	DE	2521563
STEVENS, J.	US	3527564
STEVENS, J.	US	3617220
SUDDEUTSCHE KALKSTICKSTOFFWERKE	DE	2738415
SUMITOMO CHEMICAL INDUSTRIES	EP	62496
SUMITOMO CHEMICAL INDUSTRIES	FR	2174951
SUMITOMO CHEMICAL INDUSTRIES	FR	2216227
SUMITOMO CHEMICAL INDUSTRIES	FR	2260630
SUMITOMO CHEMICAL INDUSTRIES	JP	77. 47803
SUMITOMO CHEMICAL INDUSTRIES	JP	80. 15952
SUMITOMO CHEMICAL INDUSTRIES	US	4101615

SUMITOMO CHEMICAL INDUSTRIES	US	4152149
SUMITOMO ELECTRIC INDUSTRIES	EP	56262
SUMITOMO ELECTRIC INDUSTRIES	EP	56996
SUMITOMO ELECTRIC INDUSTRIES	JP	81. 92180
SUMITOMO ELECTRIC INDUSTRIES	JP	81.100168
SUMITOMO ELECTRIC INDUSTRIES	US	3838488
SUVOROV, S.A.	GB	1511393
SUVOROV, S.A.	SU	528286
SVWA SEIKOSHA	JP	74. 40121
TAKAHASHI, T.	US	3657089
TARSHIS, L.A.	US	3783033
TEXACO	GB	1051883
TEXACO	US	3572286
TEXAS INSTRUMENTS	US	3615275
THOMPSON FIBRE GLASS	DE	1469488
THOMPSON FIBRE GLASS	GB	1030232
THOMSON CSF	DE	1944504
THOMSON CSF	FR	1505474
THOMSON CSF	FR	1535660
THOMSON CSF	FR	2036618
THOMSON CSF	FR	2165012
THOMSON CSF	GB	1190038
THOMSON HOUSTON (BRITISH)	GB	733061
TOA NENRYO	EP	44714
TOA NENRYO	EP	55024
TOHO BESLON	GB	2071702
TOHO BESLON	FR	2488244
TOHO BESLON	JA	75. 89695
TOKUSHU MUKI ZAIRYO KENKYUSHO	JP	78. 81727
TOKUSHU MUKI ZAIRYO KENKYUSHO	JP	78.103025
TOKYO SHIBAURA ELECTRIC	JP	79. 12488
TOKYO SHIBAURA ELECTRIC	US	3833389
TOMITA, C.	US	3840647
TORAY INDUSTRIES	DE	2042358
TORAY INDUSTRIES	EP	24277
TORAY INDUSTRIES	FR	2073796
TORAY INDUSTRIES	FR	2393087
TORAY INDUSTRIES	JA	72. 24977
TORAY INDUSTRIES	US	4301136
TORIKAI, E.	US	3959453
TOSHIBA CERAMICS	JP	79. 90209
TOYO BOSEKI	GB	1535471
TOYOTA	EP	45510
TOYOTA	JP	73. 55105
TRABACCO, R.	US	3847558
TRW	FR	211022
TRW	GB	1151464
TRW	US	3691623
TUMANOV	FR	2242476
TURNER, W.	US	3767773
TYCO	FR	2159338
UBE INDUSTRIES	DE	2421443
UBE INDUSTRIES	EP	21844

UBE INDUSTRIES	EP	23096
UBE INDUSTRIES	EP	30105
UBE INDUSTRIES	EP	48957
UBE INDUSTRIES	EP	51855
UBE INDUSTRIES	EP	55076
UBE UNIVERSITY	JP	81.110733
UNION CARBIDE	DE	1912465
UNION CARBIDE	EP	26647
UNION CARBIDE	EP	44761
UNION CARBIDE	FR	2123366
UNION CARBIDE	FR	2141136
UNION CARBIDE	FR	2191996
UNION CARBIDE	FR	2192067
UNION CARBIDE	FR	2192193
UNION CARBIDE	FR	2207088
UNION CARBIDE	FR	2296032
UNION CARBIDE	GB	1144033
UNION CARBIDE	GB	1159210
UNION CARBIDE	GB	1177782
UNION CARBIDE	GB	1353384
UNION CARBIDE	GB	2053177
UNION CARBIDE	US	3385915
UNION CARBIDE	US	3488151
UNION CARBIDE	US	3553820
UNION CARBIDE	US	3663182
UNION CARBIDE	US	3860529
UNION CARBIDE	US	3861947
UNION CARBIDE	US	3974264
UNION CARBIDE	US	3995024
UNION CARBIDE	US	4162301
UNION CARBIDE	US	4303631
UNITED AIRCRAFT	DE	1954480
UNITED AIRCRAFT	FR	1551091
UNITED AIRCRAFT	FR	2011507
UNITED AIRCRAFT	FR	2033230
UNITED AIRCRAFT	FR	2072682
UNITED AIRCRAFT	FR	2111243
UNITED AIRCRAFT	FR	2191978
UNITED AIRCRAFT	GB	1177854
UNITED AIRCRAFT	GB	1401371
UNITED AIRCRAFT	US	3427185
UNITED AIRCRAFT	US	3615277
UNITED AIRCRAFT	US	3640693
UNITED AIRCRAFT	US	3660140
UNITED AIRCRAFT	US	3679475
UNITED AIRCRAFT	US	3695916
UNITED AIRCRAFT	US	3698970
UNITED AIRCRAFT	US	3720536
UNITED AIRCRAFT	US	3772350
UNITED AIRCRAFT	US	3772429
UNITED AIRCRAFT	US	3787236
UNITED AIRCRAFT	US	3801351
UNITED AIRCRAFT	US	3811930
UNITED AIRCRAFT	US	3850689
UNITED AIRCRAFT	US	3865917
UNITED AIRCRAFT	US	3887384
UNITED AIRCRAFT	US	3887722

UNITED AIRCRAFT	US	3900626
UNITED AIRCRAFT	US	3903323
UNITED AIRCRAFT	US	3903347
UNITED AIRCRAFT	US	3914500
UNITED KINGDOM ATOMIC ENERGY AUTHORITY (UKAEA)	DE	2500082
UKAEA	FR	1578319
UKAEA	FR	1600655
UKAEA	FR	2014130
UKAEA	FR	2091204
UKAEA	GB	1172855
UKAEA	GB	1343773
UKAEA	GB	1410090
UKAEA	GB	1403862
UKAEA	GB	1403863
UKAEA	GB	1421672
UKAEA	GB	1455331
UKAEA	US	3704147
UKAEA	US	3828699
UNITED TECHNOLOGIES	FR	2338332
UNITED TECHNOLOGIES	FR	2374426
UNITED TECHNOLOGIES	FR	2475534
UNITED TECHNOLOGIES	GB	2075490
UNITED TECHNOLOGIES	US	3953647
UNITED TECHNOLOGIES	US	4064207
UNITED TECHNOLOGIES	US	4103063
UNITED TECHNOLOGIES	US	4148671
UNITED TECHNOLOGIES	US	4263367
UNITED TECHNOLOGIES	US	4265968
UNITED TECHNOLOGIES	US	4268562
UNITED TECHNOLOGIES	US	4341826
UNIVERSAL OIL	US	3614809
UNIVERSAL OIL	US	3632709
UNIVERSAL OIL	US	3652749
UNIVERSAL OIL	US	3814782
UNIVERSAL OIL	US	4250131
US ATOMIC ENERGY COMMISSION (USAEC)	DE	1961303
USAEC	US	3607672
USAEC	US	3671385
USAEC	US	3700535
USAEC	US	3718720
USAEC	US	3779716
US BORAX	NL	7408602
US COMPOSITES	GB	1215800
US COMPOSITES	FR	2057466
US DEPARTMENT OF ENERGY	DE	2947917
US DEPARTMENT OF ENERGY	GB	2055356
US DEPARTMENT OF ENERGY	US	3793204
US DEPARTMENT OF ENERGY	US	4152381
US DEPARTMENT OF ENERGY	US	4152482
US DEPARTMENT OF ENERGY	US	4180428
US SECRETARY OF THE AIR FORCE	US	3384578
US SECRETARY OF THE AIR FORCE	US	3386918
US SECRETARY OF THE AIR FORCE	US	3416944
US SECRETARY OF THE AIR FORCE	US	4201611
US SECRETARY OF THE ARMY	US	3550213
US SECRETARY OF THE ARMY	US	3794551
UVAROV, L.A.	US	3496078

VAN DEN BOOGAARD, J.	US	3785805
VERSAR	US	4104395
VOGEL, F.	DE	2537272
WACKER CHEMIE	GB	1177739
WAGNER, R.S.	US	3493431
WAGNER, R.S.	US	3505127
WALTER, J.L.	US	3782928
WALTER, J.L.	US	4058415
WARREN CONSULTANTS	US	4275095
WEETON, J.	US	3337295
WEIL, W.M.	GB	1108659
WESTINGHOUSE CANADA	US	4107352
WHITE, E.F.T.	EP	7693
WUERTTEM BERGISCHE METALLWARENFABRIK	US	3233985
XEROX	US	3505177

Subject Index